SCIENCE: The Glorious Entertainment

Books by Jacques Barzun

History

THE FRENCH RACE, 1932
RACE, A STUDY IN SUPERSTITION, 1937; (*Torchbooks, 1964*)
DARWIN, MARX, WAGNER, 1941; (*Anchor Books, 1958*)
CLASSIC, ROMANTIC, AND MODERN, 1943; (*Anchor Books, 1961*)
BERLIOZ AND THE ROMANTIC CENTURY, 1950; (*Meridian Books, 1956*)
THE MODERN RESEARCHER (with Henry F. Graff), 1957; (*Harbinger Books, 1962*)

Criticism

OF HUMAN FREEDOM, 1939; (*Rev. Ed. Lippincott, 1964*)
TEACHER IN AMERICA, 1945; (*Anchor Books, 1954*)
GOD'S COUNTRY AND MINE, 1954; (*Vintage Books, 1959*)
MUSIC IN AMERICAN LIFE, 1956; (*Midland Books, 1962*)
THE ENERGIES OF ART, 1956; (*Vintage Books, 1962*)
JOHN JAY CHAPMAN: WRITINGS, 1957; (*Anchor Books, 1959*)
THE HOUSE OF INTELLECT, 1959; (*Torchbooks, 1961*)
SCIENCE: THE GLORIOUS ENTERTAINMENT, 1964

Translations

BECQUE'S LA PARISIENNE, 1949
PLEASURE OF MUSIC, 1951; (*Campus Books, 1960*)
DIDEROT'S RAMEAU'S NEPHEW, 1952; (*Anchor Books, 1956*)
MIRBEAU'S THE EPIDEMIC, 1952
MUSSET'S FANTASIO, 1955
FLAUBERT'S DICTIONARY OF ACCEPTED IDEAS, 1954
NEW LETTERS OF BERLIOZ, 1954
BERLIOZ'S EVENINGS WITH THE ORCHESTRA, 1956
FOUR PLAYS BY COURTELINE, 1958-1960
BEAUMARCHAIS' FIGARO'S MARRIAGE, 1961

SCIENCE:

THE GLORIOUS ENTERTAINMENT

JACQUES BARZUN

HARPER & ROW, PUBLISHERS
New York, Evanston, and London

FIRST EDITION

LIBRARY OF CONGRESS CATALOG CARD NUMBER: 62-14520

B-O

TO

CASS CANFIELD

in friendship and

admiration

Contents

Acknowledgments

Between March 1960, when I had the privilege of delivering the Marfleet Lectures at the University of Toronto, and May 1963, when I gave a series of lectures at the University of Cambridge, I have taken every opportunity of developing in public the ideas set forth in this book. Besides the advantage of repeated recasting, I have thus had the benefit of hearing my hypotheses challenged and discussed.

Among the American audiences to which I am indebted for this help are those I addressed at the University of California in Berkeley and in Los Angeles, at Stanford University, the Aspen Institute for Humanistic Studies, the University of Colorado, Antioch College, Vassar College, the Industrial Indemnity Company of San Francisco, the Connecticut General Life Insurance Company, the Physics Teachers' Summer Institute at Columbia University, the History of Science Society, and the Sigma Xi Society at the College of Physicians and Surgeons.

Some of the passages from those talks which have survived as far as the present text had an intermediate existence in published articles. For permission to reproduce those paragraphs I thank the editors of *Harper's*, *Holiday*, *The Saturday Evening Post*, *Science*, and *The Antioch Review*. To the *New York Times*, that daily encyclopedia of the world's wisdom and folly, I owe a special debt for the copious evidence it has furnished me about contemporary action and belief.

My friends and colleagues have taken in this book an interest which has deeply touched me and for which I am proportionably grateful. It is thanks to them that the work is less imperfect than it otherwise would be. Two of these friends, Polykarp Kusch and

ix

the late Francis J. Ryan, allowed me to attend their seminar on Radiation and Human Life and pursued in conversation the subjects there broached. I further imposed on Polykarp Kusch by asking him to read chapters of the book, and others were similarly laid under contribution, notably: W. H. Auden, Daniel Bell, Justus Buchler, Theodore Caplow, Alburey Castell, Erwin Chargaff, Richard Franko Goldman, Ralph S. Halford, Banesh Hoffmann, Joseph Wood Krutch, Morris H. Philipson, Herbert Robbins, Fritz Stern, Wendell H. Taylor, and Lionel and Diana Trilling. To them all I give heartfelt thanks.

Lastly, in the preparation of the typescript, notes, and proofs, I gratefully relied on the help of my faithful assistants: Violet Serwin, Judith Gustafson, and Sarah Faunce. Virginia Xanthos Faggi made the index.

J.B.

SCIENCE: The Glorious Entertainment

I

To the Reader

THE handiwork of science in the modern world is a topic of everyday reference, and among those who lead opinion the reference is generally adverse. During the past thirty years the articulate have unceasingly cried out against the tyranny of scientific thought, the oppression of machinery, the hegemony of things, the dehumanization brought about by the sway of number and quantity. On these themes the nations most renowned for science and industry have applauded the wounded poets, the philosophers of myth, the devotees of agrarianism, the novelists of the life of instinct, and the painters and dramatists who scorn civilization—to say nothing of the day-to-day critics of mass culture.

At the same time the leaders and the popularizers of science and technology have not been silent, and in serving their more limited aims they too have managed to establish about science some widespread convictions. Together, these two choruses have supplied journalism and conversation with a number of thought-clichés. The reading and listening public "knows," among other things:

that in our culture there are two cultures, composed of the men of science and the men of non-science—an alarming split when we consider how fast knowledge is growing among the men of science;

that there has been a scientific revolution, or perhaps two—or more—for the surer saddening of man: first Newton removed God from the center of the universe, then Darwin removed Man, and finally Freud removed Mind;

that thanks to scientific analysis nothing is any longer what it seems, everything being the effect of some hidden, blind, irresistible force beyond control—except by science;

that whereas the physical sciences have made great progress and given us control over nature, the moral and social sciences have lagged behind and failed to give us control over human nature, which explains war and juvenile delinquency;

that science being the only reliable enterprise of reason, it will gradually resolve men's doubts and difficulties and thus remain sole master of thought and queen of the curriculum;

that science is incurably difficult for the layman, who ought not to compete with the few who are "mathematical" and destined for it; scientific revolutions occur so frequently that any recollection of one's scientific studies is almost sure to be wrong; and yet the new knowledge permits a return to the belief in God and immortality;

that unless science is taught to everybody, thoroughly and soon, Russia will conquer and bury us; the East, in fact, is bound to win, because it is training nothing but technicians;

that no matter what anybody does, science and technology are reshaping human life in the same way for East and West;

that technology, the offspring of science, is the cause of the century's troubles—isolation, alienation, anomie, the low estimate of art, and the breakup of the family;

that in a rapidly changing world full of technical problems, scientists should have a voice in government; although it is also true that the patient work of science unfits a man for making decisions and tolerating the hit-and-miss processes of democracy;

that the machine civilization produced by science is lifting the masses everywhere from the bondage of disease and superstition and into the higher preoccupations of the good life, such as leisure-time reading and community work;

that the machine civilization produced by science destroys all human values—they flourish far better in primitive societies. The machine breeds only urban materialists who now stand on the brink of mutual destruction by man's highest achievement, the Bomb; or, alternatively,

that the technologically developed countries stand at the threshold (not brink) of the atomic age, when the marvels of science will be widely diffused and appreciated—atomic power plentiful and clean, life made in a test tube and prolonged indefinitely on a diet of plankton, art for all, the scientific control of man's evil impulses, and the adventure of exploring, perhaps colonizing, Space.

These propositions are used to define our present (exciting) (miserable) condition. What are they worth? Or rather, which of them—if any—should survive comparison with its neighbor or with recorded fact? It is not true, for example, that technology is the offspring of science, and that art today is held in low esteem. Even apart from truth, it is regrettable that informed opinion should accept such incoherent temptations to belief: science for everybody and for the few, science dominant among us and in danger of neglect, science man's highest achievement and spiritual blight, science the benefactor and the common enemy.

Below this confusion lie many unresolved questions. It is not clear to anyone, least of all the practitioners, how science and technology in their headlong course do or should influence ethics and law, education and government, art and social philosophy, religion and the life of the affections. Yet science is an all-pervasive energy, for it is at once a mode of thought, a source of strong emotion, and a faith as fanatical as any in history.

This cultural sovereignty of science and its multiplying effects have long engaged my attention. In my previous book, *The House of Intellect*, I indicated some of the conflicts between science and the proper work of the mind. But that was not my first incursion into the subject. Twenty years ago, in *Darwin, Marx, Wagner*, I tried to describe the tradition of scientism, and even earlier, in *Race: a Study in Modern Superstition*, I examined in historical detail one application of the scientific outlook to a presumed social reality. The present book pursues the same study of cultural history and also develops some of the ideas in *The House of Intellect*. When fitted together these narratives will make clear that for my inquiry I have defined our time as the period since 1750, and that I have understood cultural history to mean the evolution of ideas, of the arts, and of social forms and feelings.

If I recall here those earlier essays, it is not only because they testify to some length of meditation on my present subject, but also, and mainly, because they bring me to mention the work which first opened to my generation the question of science as a cultural force, and which incidentally helped determine my choice of studies. I refer to Whitehead's *Science and the Modern World.* In the nearly forty years since its publication many books have appeared on the logic, the philosophy, and the sociology of science. Earlier there had been studies of the structure of representative sciences, notably the works of Stallo and Norman Campbell. But no one before Whitehead or after tried to reconstruct the situation of science and portray its states of mind. Certainly no one, whatever the aspect of science discussed, rivaled Whitehead's deliberate amplitude; I mean by this that no one saw why it might be necessary to discuss close together and with equal seriousness the nature of mathematics and the poems of Wordsworth and Shelley; to point out the medieval strain in the modern scientific view of nature; to detect in the uninflected modern languages the mark of modern self-consciousness; to characterize Voltaire as a cultural disinfectant and William James as an "adorable genius"—in short, to be historian, biographer, and critic while discussing science.

So remarkable a work, read in my nineteenth year, had upon me the effect of a vision. And if I permit myself another word of personal recollection, it is but to make plain how indebted I am— how indebted every student of western civilization must be— to Whitehead's great essay. It is true that I had been prepared for its message by an eager absorption of the works of William James, who remains the fountainhead of understanding for the student of the transition from nineteenth to twentieth-century thought. But I might easily have missed Whitehead's book so early in its life and in mine, if it had not been for the chance of being asked, by way of elevated hackwork, to summarize it for *Keller's Digest of Books.* There, I am afraid, my feeble exposition still stands, for the perilous convenience of college students in a hurry. I can only hope that my present confession, given in the year of Whitehead's centenary, may make up just a little for the youthful presumption of paraphrasing the definitive.

But if Whitehead's work was definitive and, as I think, still tells the fundamental truth about science in the modern world, why should someone without his many and extraordinary qualifications undertake today another review of the same situation and state of mind? The answer is that the situation is now more cruelly plain than before, and the evidence about the state of mind is so changed as to make its earlier phases unrecognizable.

Writing shortly after the first war, Whitehead could say that the union in science "of passionate interest in the detailed facts with equal devotion to abstract generalisation" had produced a "balance of mind" now part of "the tradition which infects cultivated thought." Today the sense of that balance is gone. What is felt is the *curse* of abstraction, the *burden* of pullulating fact. The weight of all the abstract art and thought, all the symbol-making, all the psychologizing, all the resistance to the obvious and denunciation of the standardized, all the searching for the "really real,"—in short, all the expressions of private and public anxieties felt during the past thirty-five years have destroyed the equilibrium and the confidence that went with it. The elements remain, but scattered, ill-assorted, in little heaps that give off not generalization but clichés.

Science itself, through its increased prosperity and fact-breeding specialism, has contributed to the disturbance and distress, while the technology of automation, nerve gas, germ warfare, and atomic destruction have driven home the lesson of human helplessness directly proportional to human "control over nature." The result is that whereas Whitehead defines his theme as "the energising of a state of mind," I am forced to say that mine is the debilitation of that state and the disaffection from it. Were the scientific nations of the West really in possession of life-giving truths, widely felt and accurately known, these nations would surely be seen to be happy and proud. Instead, they proclaim with conviction and unanimity their misery, their hatred of the times and of themselves. Whatever contrary boasts they may utter are conventional, perfunctory, absent-minded. Nothing in the tone of discourse on public affairs betrays a balanced and rightly energized state of mind.

Whitehead, to be sure, pointed out a "radical inconsistency at the basis of modern thought"—that between the mechanistic outlook of science and the continuing belief in "the world of men as self-determining organisms." This discrepancy remains fundamental—and with it his book. But he saw the conflict as "distracting" and "enfeebling" merely; and he offered his own organismic philosophy as the beginning of a remedy. He could not foretell that long before any such cure could take effect the western world would be "distracted" in a far stronger sense.

Nor did the scheme of his work allow him to pursue the difficulty elsewhere than in the realm of philosophy and religion. That, to the best of my knowledge, is where it has been left; and it is from that plane of abstraction that I wish to remove it in this book. No one, I believe, has bothered to trace the ways in which the modern oppression works. The enfeeblement and distraction have been felt and diagnosed rather in generalities than in particulars, because for those who live among them the particulars seem too obvious to mention. And this again leads to false explanations and stereotypes. The complaints being sweeping rather than exact, they call for the total abolition of the social order rather than individual resistance to its evils. I should except Mr. David Riesman's *Individualism Reconsidered*, which offers well-reasoned prescriptions for a saving autonomy of the personality within the mass world, but these suggestions proceed as usual from a "scientific" base, which I believe ultimately untenable.

From this remark some may hastily infer that I am "against science." And indeed this charge was occasionally made by readers of *The House of Intellect*, despite the fact that it contained a parallel discussion of art, which led other readers to say that I was "attacking art." Such notions of attacking and being against are perhaps native to the age; they are foreign to me. I do not understand what is meant by being against, or for, wholes—art, science, education, medicine, the state. I can as readily imagine being against sunsets and for the tides. Criticism as I understand it differs entirely from attack or complaint. Its difference from complaint is especially important here, for I am persuaded that complaints against the machinations of culture today have become as poisonous as the things complained of. This is not surprising. Resentment and indignation are feelings dangerous to the possessor and to be sparingly used. They give comfort too cheaply; they rot judgment, and by encouraging passivity they come to require that evil continue for the sake of the grievance to be enjoyed.

Criticism, on the contrary, aims at action. True, not all objects can be acted on at once, and many will not be reshaped according to desire; but thought is plastic and within our control, and thought is a form of action. To come to see, in the light of criticism, a situation as different from what it seemed to be, is to have accomplished an important act. The contemporary world, cluttered with leagues and lobbies and overawed by the zero-weighted look of large numbers, has forgotten that to redirect fundamental opinion—including one's own—is also to *do* something. It can give solace or mastery, or at the very least replace a plaintive passivity with a stoic impassivity.

That is why, in the present work, which is primarily descriptive, and which, like its predecessors, often uses for illustration not what is merely typical, but what is especially worthy of regard, I sometimes employ the rhetoric of argument—the form that most naturally incites the internal action called thought. To the reader it should not greatly matter whether or not he agrees with the conclusions I reach. The point of offering them is to reduce confusion and provide a spur to reasoning. For the aim of a critic, beyond that of saying what he thinks, is to make two thoughts grow where only one grew before.

Such an aim implies a reader willing to follow discourse even though the subjects brought together are very diverse, and to lend his mind to attitudes that may seem dubious coming from one who lays no claim to being an expert. But in matters of public interest the public must judge and not ask its guides for certificates and guaranties, for there are no guaranties. It is an error to suppose that when a physicist talks about science he is bound to be more reliable than a so-called layman who has taken the trouble to inform himself and to think. For if scientific specialization means anything, it means that the physicist has final authority only on such questions of physics as fall within his specialty.* Whitehead, whose

* "The intelligent public," as I. I. Rabi has said apropos of science and secrecy, "[is] quite capable of judging. . . . Less inhibited men of greater or lesser scientific

eminence was first as a mathematician, devoted his life to grappling with questions of philosophy, history, literature, and education. These are public questions, and he served the public interest by bringing to bear upon them his vast intelligence and experience of life. So can anyone serve in the degree of his ability, provided he uses his intellect as a guide in the great regions outside his narrow profession.

No one denies that the problems which special methods solve within the professions are important, but they affect the conduct of life only insofar as the solutions are caught up in common ideas. How we get to the moon is largely (not wholly) a technical problem; whether we want to go and what we do after we get there are public questions on which any reasonable voice should be heard. And reason of course implies preparation, without on that account excluding error.

To guard against error, rather than to humor the prejudice in favor of expert authority, I have asked—as is the academic custom —a number of colleagues to read my manuscript. Half of them are scientists, and of these several have won great awards. They naturally do not all agree with everything I say, nor do they agree with one another about what they accept and reject in my views. But they concur in believing that nothing directly deriving from scientific work is misstated in these pages. I mention this, not to escape criticism, but to caution the scientific reader who might be tempted to take his divergent understanding of certain matters as being universally "held by science." The "unanimity" of the scientific world is not what many honest workers in the field suppose —and that too is a fact of our scientific culture.

---

or technical accomplishment, but with a low boiling point, have been gaining the public ear on the basis of prestige acquired through a technical accomplishment quite limited in scope. . . . The fear of being guilty of judgment based on a partial knowledge of the facts misleads many judicious people into accepting judgments by others whose knowledge is often even more partial. . . ." *Atlantic*, August 1960, 41-42.

II

One Culture, Not Two

To USE the word Culture in the meaning it bears for the educated reader today is to enter at once into the scientific state of mind. For culture in that sense is not learning or cultivation or even the local society one lives in: it is an object for the inquisitive mind of man to study and report on, an object considered as if it existed by itself.

You may accordingly like to think of culture—I often do—as an enormous pumpkin, hard to penetrate, full of uncharted hollows and recesses for cultural critics to get lost in, and stuffed with seeds of uncertain contents and destiny. Or again, it is an interminable marine monster, like the one at Loch Ness, dimly seen and badly photographed, about which endless debate goes on. Though many have observed its twistings and some have been shaken by its commotions, no one has cast his eye down the whole length of it and given a complete description. One reason may be that we are all cramped and darkling inside it, like Jonah; but we are none the less determined that it is an object out there, acting on us, apart from us, and of which, despite its size and obscurity, we can find the secret. Culture is thus like the universe as science conceives it.

This object-like view of culture is a recent acquisition. It dates from about seventy-five years ago, when social questions were reclassified and renamed on the pattern of the natural sciences. Anthropology was one of the subjects redefined in this intellectual renovation at the turn of the century. Concerned with the varieties of mankind and therefore studying the primitive, anthropology took as its scientific unit the whole society—all the mental and material workings of a people. This it called "a culture." From looking with detachment at every word and gesture of a self-contained group,

9

and from the envious pleasure that western man took in the reports of growing up in Samoa or elsewhere, came, by imitation, the habit of looking at ourselves anthropologically, as "a culture."*

The first consequence of the new attitude for literature and common thought was an increase of self-consciousness, which proved at once liberating and crippling. Thoughtful men lost confidence in their own customs, but in compensation cultural criticism became the playground of fancy, which freely disported itself, at home in the unprovable. Soon, like other popularized technicalities, the anthropological method became a trick of the sophisticated mind and finally a mode of everyday speech. It is now a journalistic commonplace to write: "In a scientific culture such as ours. . . ." The anthropologist's view of culture as a natural object becomes for the public a scientific fact, and that object is an agent: "Culture brings about" this or that; "our culture produces . . . determines . . . demands. . . ."

The vagueness of such formulas is convenient in a democracy when one must distinguish social groups. Few are sure of being understood if they speak of classes, and "sect" sounds old-fashioned and inappropriate. The talk is of cultures and subcultures, thanks to which it is possible to express curiosity, animus, or alarm. That is how, in the current preoccupation with science and technology and the future of the West, one of the best-informed observers of the total scene, C. P. Snow, ventured to propose—and has all but established—the diagnosis of "the two cultures."†

He meant by this that a main feature of our society is the split of its intellectual men into two parties having little to say to each other. The difference in language and cast of mind between men trained in the sciences and men trained in the humanities seems to Sir Charles so radical as to justify his calling each group a culture:

---

* The tendentious comparison of primitive and civilized life goes back of course much farther than the 1890s. Tacitus in the first century A.D. was perhaps a *late* example in western civilization. In modern times, Montaigne and the eighteenth-century philosophers used the same method to establish ethical relativism; Rousseau changed the face of Europe by his cultural criticism, and Balzac in his Foreword to the *Comédie Humaine* (1842) anticipated the behavioral scientists in representing his work as the "zoology" of contemporary society.

† This notion of the divergence between scientists and humanists is suggested in Whitehead's *Science and the Modern World* (111), but on different grounds from Snow's.

the scientists cannot read Dickens and the humanists do not understand the meaning of "mass" and "acceleration." Each type of man lives in a separate world, thinks with an exclusive set of terms and symbols, aims at goals shared only by those in his group, and thrives in complacent ignorance of the ways and aims of the other group— as would the inhabitant of a distinct culture. Being a novelist, a physicist, and a civil servant who travels over the entire ground of our own unified culture and, as such, feeling the urgency of problems scientific, literary, and political, Sir Charles is concerned about the ill-effects of the split which he describes.†

The crack will obviously not repair itself. If one talks first to a representative scientist and then to an art critic or literary man, their remarks suggest that they are rapidly moving in opposite directions. But this is probably an illusion, as I shall try to show; and in any case it does not mean that two cultures confront each other across the widening chasm. To begin with, neither of the two groups is homogeneous: persons and interests intermingle. What is more, when added together the two "cultures" leave out ninety-nine-hundredths of the population. Does this remnant form a third culture, neither scientific nor humanistic?

Besides, within each of these two clans precision would require that we distinguish subgroups, as isolated as the parent pair. Consider the meaning of "avant-garde" and "academic" in the arts: these terms mean that in any given art parties exist which refuse to acknowledge kinship. They are thrown back on mutual scorn as the only bond they can find. And within the subgroups radical divisions continue. The same is true within science. Radio astronomers and observational astronomers, algebraic topologists and ordinary analysts, decline to speak the same language even when they can. Their supposed common devotion to science, their presumed inability to read Dickens, do not make them fellow members of one culture, clearly distinguishable from that of the Action painters and the Old

† Since these words were written, a violent attack by F. R. Leavis on Snow's person and position precipitated a noisy controversy in the leading journals of England and the United States. Both sides showed by their anger and incoherence that they considered the issue deep and moving, but they brought to it (barring one or two participants) little that was original. Snow himself maintained a proper silence.

Dickensians, who likewise address each other rarely.

The truth is that all these men, together with those for whom the issue does not even begin to exist, belong to one culture, the scientific culture of the western world in the twentieth century. If there were not in fact one culture, where would be the danger of occupational castes? If modern culture were not one, why did the discussion of the split bring forth so many plans for teaching science universally and the humanities more effectively? The answer is: Tradition. We have been reared, we live, and we believe in one culture, a culture which is common and traversable despite its compartments. That culture is now properly called scientific, because its defining characteristics come from science; or rather, it has grown scientific along with what we artificially set apart as Science.

To determine in what sense, other than casual, this is true requires an understanding of science not usually found among either humanists or scientists. Science is an awe-inspiring force, known to do remarkable things. Everyone, therefore, is presumed to know also how and with what effect science works. But few in fact pay attention to the nature and consequences of an activity so deeply approved. When this state of affairs obtains in a society about its dominant intellectual pursuit, the gap between popular reverence and common lack of knowledge is filled by conventional beliefs that range from clichés to superstitions. Gradually the relation between men and their own acts is hidden by the conventions, and one has to wait for the future historian to say what went on. There is thus something premature about my attempt to describe the chief workings of the scientific culture in which we live, for I am bound to be blinded by certain conventions too. But it is less important at this time to be right than to try to see effects and connections that are difficult to trace or that we have stopped noticing.

It is not clear, to begin with, what we mean when we speak of science. Is science a method of inquiry, the result of the method, the application of the results, or—which is very different—the understanding of the cosmos vouchsafed by the method and its results?

And a second large question that needs settling is: Who shall answer those previous ones? It is not at once apparent who is qualified to be the interpreter of a society in which science is so near and yet so far. Are the scientists alone to judge their role, and if so, which scientists among the innumerable kinds there are? Or should we go to the sociologists, who have assumed the privilege of the very young to poke into every corner? Perhaps we should consult those ex-scientists who have gained worldly experience and a public position? or the government servants and university administrators who have studied the scientist in captivity? or as a last resort, the cultural critics, who are almost as numerous as the totality of articulate citizens?

Supposing that we knew where to turn and were given the right definitions, we should still face the riddles posed by the current cant about science. For example, what does a respected atomic physicist and leader of engineering education mean when he asserts that "from now on science and technology will shape the world"? The phrase is not particularly his; it must occur a hundred times a day in the speeches of political and business leaders. The assumption seems to be that science and technology work by themselves, but we are not told what "world" they will change. It is not likely that an enthusiast for science is thinking of the odd shape the human world will present after an atomic war, though more and more people tend to associate science with destruction. Others have utopian visions of ease and plenty resulting from applied science. In either case, the argument takes for granted an inevitable future, the result not of human wills and actions, but of the acquired momentum of science and technology. This seems to invite scrutiny, but by whom?

Again, public opinion has passively accepted the belief in a "scientific revolution," which is said to have recently taken place or to be still going on. Given the modern abuse of the word "revolution," which often means no more than a change of names and slogans, it is hard to define the supposed revolution. Is it a physical or a mental change? Who is gaining power and who is losing it? Does the belief assume the identity of science and technology and simply point to rapid mechanical progress? Is the revolution the shift to the study of energy in place of matter, or does the term refer to a social movement, by which science will monopolize all talent and re-express all knowledge in mathematical form? To test the reality of these beliefs would be a laborious study in common opinion. But these unresolved ambiguities are symptoms of our state, and I cite them to show the encyclopedic carelessness with which this century speaks of its proudest achievement.

Let us then try to state what modern science is. Since the living soul of science is not easily accessible, as we shall see, and since free discussion runs the risk of being misinterpreted as hostility, I begin my answer with a eulogy: I affirm without irony or reservation that modern science is one of the most stupendous and unexpected triumphs of the human mind, just as modern machinery is a collection of breath-taking artifacts often too complacently received by a heedless, ignorant public.

But praise is not definition, and true definition shows things in their purity, in a perfection which is bound to be dimmed in the world of practice. Still, a definition is needed: I understand by modern science the body of rules, instruments, theorems, observations, and conceptions with the aid of which man manipulates physical nature in order to grasp its workings. Taken together, these ideas, symbols, and apparatus form the subject and method of the so-called pure sciences. The adjective scientific is thus capable of a strict meaning: it is properly applied to any of the enumerated elements. When applied to a brand of toothpowder or a method of piano playing, the word loses its force; it suggests only: "based on some conclusion of one of the sciences"; or more often merely: "tinged with one of the externals of scientific work." And sometimes the word degenerates into a vague honorific, synonymous with the advertiser's "reliable" or "guaranteed."

When one speaks of modern culture as scientific, neither the strict nor the loose meanings are intended. What is meant is that the elements of natural science exert on common life a prevailing influence, impart to all activity and thought a characteristic coloring. A scientific culture is a type of contemporary society through which the ideas and products of science have filtered—unevenly, incompletely, with results good and bad, but so marked that they affect the whole unmistakably. The test is simple: not only does such a society support and respect many varieties of scientific workers, but these are so placed that their work gives off emanations, practical and spiritual, which affect every part of the culture.

Here it may be objected that the emanations are not part of science; they are unintended by-products for which science itself must not be blamed. Errors, disasters, pseudo-sciences, superstitions, have no connection with the noble essence of science; they are the doings of an ignorant and imitative society. To this the reply is that blame has no place in the intended description; but that if description is to be exact, the so-called side-effects cannot be excluded from the account. The culture is scientific because science pervades it. Science is not an isolated hobby, but an overarching ideal; and like every other ideal it cannot realize itself in a vacuum. Supposing the method and principles of science were set down in a book and never tried out, its greatness and beauty would be, not increased, but diminished. The purpose must be acted out, and thereby distorted, before it can claim our full devotion. In this regard science resembles all great and good things: the only kind of science we know or shall know is one that has to work in the midst of a credulous mankind, poorly taught, ruled by mixed moralities, and bent upon many other things than the search for truth. The public which receives and distorts science is no different from the public which receives and distorts art, and it is foolish to expect a better fate for either in the future.

The scientific community itself, consisting of specialists who are also men and citizens, shows the same imperfections. Human diversity and indeed mere numbers affect every intellectual pursuit, and it would be difficult to ascertain at a given moment from "the scientists" as a group what the latest deliverance of "Science" was on any given point. We should find only the usual confusion, uncertainty, and debate. Max Planck tells us in his *Autobiography* that he does not believe that science progresses by the self-correcting action of minds. Rather, he thinks, the new ideas are promoted by the young as being new and especially theirs, and the old men with the wrong ideas eventually die off.*[1] This is undoubtedly the truth, and from its pathos I draw a further argument in praise of science. For it has found the means, despite human frailty, to make its findings largely cumulative and coherent, and thanks to its control of

* Superior numbers refer to a section of notes beginning on page 309.

terms, reasonably stable and clear.

No one, then, should feel upset because a scientific culture is a hodgepodge which owes its character as much to the derivatives of science as to the rigorous core. A scientific culture is one in which the structure of the atom is studied and comes to be more or less vaguely known; it is also a culture in which a winning racehorse is named Correlation and in which the matching pieces of a woman's suit are sold as "coordinates"; it is a culture in which the photograph of the diffraction of electrons by zinc oxide resembles a prevailing mode of painting, and in which a polished roller bearing or a magnified diagram of the male hormone, when colored, can be used as a decoration in the home.

Even without these tokens, no impartial observer of scientific culture can deny its remarkable unity and range. A parallel that may help to describe it is medieval culture in the West. Though often called a religious culture, it is more exact to call it theological, in the same fashion as ours is scientific. Just as but few men were then theologians, so few among us are scientists. Just as the language of theology ended by permeating common thought, so with us the language of science. The ultimate appeal was then to the certitudes or sanctions of theology, as with us to the certitudes or sanctions of science. The highest concern of the culture was to support, to perfect, and to disseminate theological truth and practice, as with us to support, perfect, and disseminate scientific truth and practice.

This is not to say, as some have done, that we "worship" science —the phrase is a careless metaphor; nor does the medieval parallel imply an exact correspondence between two intellectual enterprises, even though, as we shall see, medieval abstraction and analysis are closer to science than science or scholastic thought is to the pragmatism of the artist and the historian. It is worth noting that "layman" now means "not a scientist" as before "not a cleric," which suggests the same faith in the one kind of knowledge worth having. The similarity between science and theology obtains again in the crude, fitful, uncontrollable relation between a dominant intellectual system and a large heterogeneous society, which does not so much assimilate the thought as take from it a tone and a direction.

So regarded, our present western world shows, I repeat, not two cultures but one. The very conception of "a culture" means that it is one and indivisible. If it shows a mixture of interests and tendencies, that fact merely defines a large and well-developed type of culture. Our civilization has reached the stage where groups coexist without sharing the same concerns, or without recognizing the same symbols even when they share the same concerns. Those who understand hydrogen bonding do not understand insurer's bonding, and the general public understands neither.* But the widespread alarm over the increasing diversity in the languages and aims of intellect is further evidence that people today still want to be at one through cultivating the sciences, because science, like theology before it, is the reigning queen.

With these definitions and parallels in mind, our present situation seems, if not simple, simpler: our Culture is one by reason of its common roots, its long history of exchange and assimilation, and also—and chiefly—by the universal trust it has acquired in science and scientific method. The scientific stance is everywhere, even among the overt enemies of science; it is the strongest unifying force, because in the world of thought it is the only one. True, art has latterly won from the western masses a naïve adoration, but as I shall show, art too is under the sway of science, which it resembles in being a pure and self-justifying activity. That is why today these two alone, art and science, are revered, however differently, and act proud, sensitive, and resentful of challenge. While business and the state are scorned as frauds and religion is dismissed as a private matter; while the reality of love is sneered at and life itself is often repudiated, science and art are held above reproach as disciplines a man can worthily give his life to and be fulfilled by.

But art is the lesser power. Men have turned to science because it seems to give truth an exact and inalterable embodiment, and in a sense this is no illusion. But in another sense, science alters much more than art and the humanities, which according to cliché have

---

* If one wants to receive the impression of a truly alien group in the midst of familiar life, one should read about the habits of the blind, in whose lives radically different assumptions are at work. I recommend Thomas D. Cutsforth's remarkable book, *The Blind in School and Society, American Foundation for the Blind*, New York, 1951.

not "progressed." The humanities, moreover, lack a stance of their own. Hence their tendency to ape science, which confirms once again the inescapable oneness.

The spectator may take pleasure in this uniformity, but it leaves strange questions. How explain the paradox of a society which is bound by so ardent a faith to one undertaking, yet by and large withholds its mind from it, saying it lacks the time and the talent to do anything but remain ignorant? And how is it also that the sacrosanct enterprise, without being formally criticized or disparaged, has alienated the minds of those most alive to the motions of man's spirit? Here certainly we differ from the medieval world, in which clergy and clerisy were one body, however divided on points of doctrine.

The questions are of two kinds. One is historical; it asks us to account for the common vagueness about the fundamentals of science, which one would suppose to be its oldest and simplest parts: what it does, how it works, which part of it is truth and which assumption, and how far its rights extend. The second sort of question is philosophical. It wants to know about the moral and material effects of science; how the grateful wonder about antibiotics merges into the fear of the greatest antibiotic of all, the Bomb. And before the Dance of Death, there is to be explained what many ascribe to the workings of science—the relentless self-destruction of the modern ego, the misery, accidia, wanhope; in short, the conviction that mankind is at the helm of a black ship bound for hell. To have brought us to this point, to have turned our culture scientific and made it conquer the world, only to leave us in despair while still largely ignorant of our accomplishment, there must have been at work some powerful motive inspired by an idea. How did western man, reared theological for a thousand years, become "scientific"?

The best answer is that he early developed into a maker and lover of machines and has lately come to believe them products of science. The layman says "science and technology" as if they were identical twins. They have in fact distinct origins and histories, and they are still different in nature and purpose, but their union—in part a very recent development, in part a false association of ideas—has engendered the common faith in science.

Even when this link is seen, the transformation of the common mind and the common habits in less than three hundred years is a puzzle for the historian. The story is still so imperfectly known, and so absurdly represented as the clear triumph of truth over error, that one must try to re-create actual notions and sentiments with the aid of contrasts and parallels.

A man of Galileo's time, even if educated, saw no connection between the peasant's wheelbarrow and the physicist's speculations on his *Two New Sciences*. But a man of today thinks he sees the truth of science demonstrated in every gadget he uses. He thinks he grasps the goal and justification of science when he sees the increased production of food, shelter, and clothing which has come from the harnessing of modern power to modern tools.

The reasoning is obvious: the appliance works, its principle is understood, and it is made in thousands of copies with decreasing human effort. From the Power and the Device springs the great argument that has redirected the western mind. We are scientists by proxy on the strength of the work done by Newcomen and Stephenson and Watt. But science was not the initiator, it was the beneficiary, of mechanical invention.

To be sure, as far back as the thirteenth century some purely scientific ideas challenged the theologians' monopoly of explanation. But ideas alone do not change an entire culture, they require the aid of strong and simple motives. The handful of disinterested men who follow truth wherever it leads cannot act directly on the masses, and masses are moved by what is at once tangible and hopeful. We consequently ask what bound medieval man to theology as we are bound to science. Clearly, it was the desire for a permanent happiness, for a reward offsetting the pains and privations of this world. Under theology the Power and the Device to supply this want were, as with us, lodged in a class of men—the clergy—whose language differed from the vernacular and whose modes of thought were remote from the ways of common sense. But since the general longing was for a better life, the clergy's strongest competitors were not the men of science telling a different story of the heavens; the true rivals were the inventors who said nothing but increased man's power to wrest a living from nature.

If we like to fix history in our minds with symbolic events and dates, we may choose the original appearance of Dr. Faustus and the date 1587 as a hypothetical turning point in western culture. For in that year was published the first account of Faust's compact with the devil, an established Power and Device which an enterprising magician like Faust would naturally come to terms with. The bargain that he made then should not be confused with later versions of the tale. The favors Faust was asking for in the sixteenth century were not knowledge and a beautiful woman to love. They were: abundance of food and clothing. Faust also asked for pocket money and, by way of entertainment, he expressed a wish "to fly among the stars."

In the next two hundred years the abundance of food and clothing was sought and found by technologists, and we are getting closer to fulfilling the last request. The means to all this came from artisans and other clever men who built devices and developed power without the aid of scientific method or principle. The steam engine, the spindle frame and power loom, the locomotive, the cotton gin, the metal industries, the camera and plate, anesthesia, the telegraph and telephone, the phonograph and the electric light—these and other familiar wonders were the handiwork of men whose grasp of scientific ideas was slight, or at best empirical.

By a coincidence not wholly clear, while these inventions were being made or perfected, a succession of great mathematicians and scientists were erecting a new system of ideas and symbols, and propounding it as the scientific order which is now above cavil in the intellectual sphere. But whereas the achievements of the inventors soon became vivid realities on the public scene, the discoveries of the scientists remained by comparison hidden and hard to grasp. The original difference between technology and science thus persisted for generations.* It began to fade when men of science became interested in measuring and explaining ordinary machines and in sug-

---

* Sir John Herschel was perhaps the first eminent scientist to address the public on the theme of Science the Helper. He dealt at length in his popular *Discourse on the Study of Natural Philosophy* (1830) with the influence of science on the well-being and progress of society, implying now and then that the latter was the justification of the former. Eighty years earlier, the French *Encyclopédie* had juxtaposed science and the practical arts, but had not conceived of them as a single human effort.

gesting applications of their own newly discovered principles. Thereafter the identity of the two movements was assumed, though not without inconsistency. It is only in our time that the big industries have established research divisions, where science is practiced for the benefit of technology;† only recently and timidly that governments have come to recognize the value of fundamental research.

At this point our original question about the reign of science takes on a new form: how did mass society approach science close enough to acquire unquestioning respect for this mode of thought, which is not only difficult but also in constant flux? How did the skeptical, argumentative, irreverent fellow freed by the revolutions of 1789 and after convert his strong attachment to machinery into a blind faith in signs and conceptions he cannot grasp? At some point there must have been for the common man something like a discovery of the existence of science.

That event, if one can call it so, was late in coming. The steady progress of the scientific elite goes back three hundred years; but the scientific departments of most universities are barely seventy-five years old, and the public awakening which preceded their establishment goes back no more than a century from us. It was in the mid-eighteen-fifties that the achievements of the Age of Reason in astronomy, physics, chemistry, and biology were given sudden and violent publicity by the controversy over Darwin's *Origin of Species*. Evolution was not new and Darwin did not discover it, but the polemic about his book had the effect of bringing everybody up to date about all of science. For whatever else it proved, the debate made plain two things: One was that the men of science formed a powerful party whom western civilization would now have to reckon with. The other was that pure science was engaged in sketching, bit by bit, the plan of a machine—a gigantic machine identical with the universe. According to the vision thus unfolded, every existing thing was matter, and every piece of matter was a working part of the cosmic technology. The rules of its functioning were being rapidly

† To this day, one of the largest chemical concerns in the United States, which has had a research division for thirty-five years, continues to print a booklet explaining to its employees and stockholders why undirected research is worth supporting.

discovered and would soon be within the grasp of all who cared to give the requisite attention.

Modern science can date its cultural supremacy from the conclusion of that period of intense argument and popularization. Like a radical party, the articulate scientists appealed to the uncommitted mob of the young and the spiritually discontented, over the heads of the Establishment. For forty years adulation of science in the abstract and controversy about what it concretely implied raged throughout Europe and America. New inventions, new sciences and pseudo-sciences proliferated. Domestic and foreign policy, social and economic issues, were fought on would-be scientific grounds. Old religions recast their dogmas to "meet" those of science, and Christ himself was called a scientist. Finally it was clear that the birthpangs of the scientific culture were over. By 1900 science had conquered its share of the curriculum, won a regular place in the press and the pulpit, invaded literature and the common tongue, and begun to alarm peoples and governments with the incalculable powers it might unleash.

For the educated, the authority of science rested on the strictness of its method; for the mass, it rested on its powers of explanation. It explained by the analogy of the machine—the simplicity of the great push-pull principle, which was everywhere applicable, permitting everyone to verify laws and formulas for himself, by sight and common sense. The great propagandists, the Huxleys and Tyndalls and Spencers and John Fiskes, often said that science was but organized and extended common sense; and they may well have believed what they said in the heat of missionary zeal. For until the rise of quantum mechanics and relativity physics, science still dealt chiefly with the large and visible portions of the universe. Leading scientists were willing to affirm that for any formulation of science a mechanical model could be built. It was the high moment of popular science in the best sense, when the scientist, as yet only in part a specialist, could lead his lay pupils from the entrancing phenomena on earth to the spectacular workings of the heavens, through an ascending series of machine-like structures conforming to a handful of superbly simple general laws.

It was a high moment in civilization also, for science made the power of mind tangible as nothing else in history had done. For the first time even boorish ignorance was cowed into respect as it dimly grasped that here was no fraud or magic, no subtlety of art rebuking grosser sense, no luxury product for the rich and well-born, but on the contrary an exercise of mind democratically open to all. The common virtues of patience, neatness, accuracy, and modest ambition were all that anyone needed in order to partici-pate and even to contribute.

If science had stood still on the apparently firm base of New-ton's four laws, had only refined its measurements of conservation and atomic weights, had succeeded in subsuming electricity under mechanics and proving the existence of the ether, always increasing its power to predict until it filled out the grand system foreseen by the Galileos, Descartes, Newtons, Goethes, and Darwins, the peoples of Europe and America might shortly have mastered, if not the scientific habit of mind, at least the scientific vision, and the historian of ideas might have spoken with literal truth of a "scientific revolution."

But just when it had captured power and authority, nineteenth-century science had to face insurrection within, a radicalism sprung from the anti-mechanical facts of electricity. As these difficulties were tackled one by one and revealed the strange workings of both the subatomic and stellar worlds, the great vision broke up, the fragments became less and less graspable, specialties multiplied, the proofs by observation and common sense gave way to mathematical demonstration—in short, science lost communicability through words, and with it lost the eager and informed interest of the edu-cated, though keeping their respect. For a time the same turn of the wheel brought hope to the moralists and religious philosophers who had been repelled by the spectacle of a universe as mindless as a machine. Twentieth-century science, they were assured by some scientists, leaves room for the creative act and for religious and moral freedom.

But this relief was local and temporary. The unsophisticated public of the nineteenth century had been converted with a finality that later ideas could not upset. To the extent that one can speak today of a common view of science, it is that of nine-teenth-century mechanism tricked out with newer words.* It is the boy's view, matching his taste for popular mechanics; it domi-nates ordinary speech, and it serves most attempts to reach the layman. No newspaper reader finds it strange when man is called a chemical machine, and many scientists themselves readily equate the operations of their own minds with the work of a digital com-puter.

There has thus been no scientific revolution at large; that is, no completed revolution of the intellect caused by science. There has been rather a social revolution, which has enthroned science in the name of increased production, increased communication, increased population, and increased specialization. The burst of zeal for a comprehensive knowledge of the cosmos through science, which shook western society for a brief time in the second half of the nineteenth century, was a putsch that failed.

The failure was not due solely to the increasing difficulty of science. In its pure state it has always been a refined exercise denied to the many. But its translatability changed with the new century, and the composition of the public also. The social revolution favored by technology and democracy enormously enlarged the reading public and shaped it into a mass impossible to address and teach in the old manner. The schools turned away from the intel-lectual disciplines and became agencies for adjusting the young generations to the complexities of an industrial world. The tech-nology of so-called mass communication seemingly extended the reach of ideas, but actually dulled public attention by excess and abolished the capacity for quick, common response. In short, the gains produced by the ideal of Diffusion proved, as might have been expected, diffuse. For the first time experience showed that it is possible for new means of life to engender new difficulties with which the means cannot keep up.

To make confusion worse, the first world war caused a break in

* "The average practicing scientist is a materialist," says a student of method who is also a mathematician and astronomer. John G. Kemeny, *A Philosopher Looks at Science*, New York, 1959, 219.

the continuity of western culture which, as I have shown elsewhere, we have not yet recovered from.* Just as in art and philosophy we are only now beginning to assess correctly what was done in the first two decades of this century, so in science the new visions of those early years have been conveyed to us garbled or not at all. The new science is for the public a Delphic mystery; it keeps the western intellect troubled but unenlightened, except for the practitioners themselves. Here we touch upon the grim deficiency of the scientific culture, which is also the first general conclusion to be drawn from our historical review: the fundamental lack in our mental and spiritual lives does not come from the trifling division between scientists and humanists, or between scientists and the whole of the laity; it comes from the fact that *science and the results of science are not with us an object of contemplation.*

By common consent the sciences are taught in school and college; but everyone admits that this teaching is wasted on three fourths of those who are forced to endure it. For all the use or interest they find in "the science requirement" they might as well be required to take Greek and Latin: science remains to them a dead language. The belief persists that if the sciences were better taught they might prove more titillating, but there is little agreement on how to teach them. And one reason for this is that to the teachers and practicing scientists themselves science is rarely an object of contemplation. They teach, therefore, as if to intending professionals. Some of the best scientific minds who have given thought to the matter repeat that there is nothing to say *about* science; that science ceases to be science the moment one is not working at it; which means learning by doing it, in preparation for research and discovery. This puts science in a category by itself in the curriculum.† If this activism of science were carried out in other subjects, the young would not be taught to read and criticize literature: they would be trained to write novels and poetry; they would not read history, look at pictures, or listen to music with

* Cf. *Classic, Romantic, and Modern*, Epilogue, and *The Energies of Art*, Introduction.
† Unless one includes physical education.

suitable commentaries; they would begin as apprentice historians, painters, and composers, regardless of talent or inclination.

To say this is not to dispute what men of science think about teaching their subject. It may well be that modern science offers nothing that man as man can contemplate and discuss. But if this is so, science cannot form the basis of culture in the old sense of cultivation for pleasure and edification. By pleasure I do not mean dabbling. The acquisition of facts and principles is prerequisite to all cultivated enjoyment. But acquiring facts either for professional use or for no use at all is bound to leave the majority of the educated discomfited by science—as we find them to be. They are not neglecting a "culture" in the honorific sense, for it does not exist as such even for their scientific friends.

As a field for cultivation, literature starts with two advantages: it deals with common experience, couched usually in common language, and it has a long tradition of being looked at and talked about. The corollary is that when the man of science finds it difficult or impossible to read Dickens, the deficiency is in him or in his training; but when the humanist finds it difficult or impossible to talk about science, the deficiency is in him, in his training, and also in the present state of the subject.

There was a time when scientific knowledge formed part of every educated man's mental furniture, and it is often said that such catholic knowledge in the educated was due to the meagerness of science and the many false ideas about it. This explanation is unsatisfactory. The genuine science of that older day was doubtless meager, but those many false ideas made it a large subject, as large as any that any man now holds in his mind. What the learned know is probably a constant quantity, and the progress of science from decade to decade suggests that the proportion of carefully learned error is also constant.

What is new among the conditions of knowledge is the result of what is usually called specialization, but should rather be called specialism. Specialization—attending to one thing at a time, and for as long a time as will insure thoroughness—is obviously desirable and it is not a modern invention. Specialism is something else: it is a piece of etiquette which decrees that no specialist shall bother with the concerns of another, lest he be thought intruding and be shown up as ignorant. Specialism is born of what the philosopher Arthur Balfour called "the pernicious doctrine that superficial knowledge is worse than no knowledge at all."[2] A little learning is dangerous in one who tries to teach or use that little in professional work; it is not a danger but a source of pleasure to the observer of life as a whole. Thus does a map, yielding a superficial knowledge of geography, add to the traveler's enjoyment even though he himself could not survey the ground and draw the map.

This failure to distinguish the concern of the maker from that of the observer is a characteristic of the scientific culture and another tribute to the force of science within it. Because division of labor in scientific work proved useful, recognizing proprietary rights in specialties became a point of honor everywhere, even where it served no purpose but to curtail pleasure. Every subject today confronts the tyranny of the professional, with the result to be expected: the reduction of every art and mode of thought to a preoccupation with the details of its making. In parallel with science, these formerly public goods are being withdrawn one by one from among the objects of contemplation; that is, of direct, unscholarly, unpedantic enjoyment, discussion, and criticism. True, the so-called specialization is frequently decried, but in the tone of hushed hopelessness, as an enslaved people might speak of the age of heroes. The rampant specialism, an arbitrary and purely social evil, is not recognized for the crabbed guild spirit that it is, and few are bold enough to say that carving out a small domain and exhausting its soil affords as much chance for protected irresponsibility as for scientific thoroughness.*

So far, then, from facing a two-party system in the house of intellect, we face an endless separatism based on loyalty to the parish pump and linked with the unanimous desire to obliterate the mere observer by denying the validity of his role. Everyone must be a maker in his own corner and take pride in his exclusivism. The dogma prevails that only the man engaged in doing

* On the subject of professionalism I should like to draw attention to a remarkable work, wise and passionate, and also entertaining, which has been neglected since its appearance forty-seven years ago: Frank H. Hayward's *Professionalism and Originality*, London, 1917.

is entitled to open his mouth. But when applied to he naturally refuses to do so, on the indisputable ground that no one is prepared to understand him.

The plain effect of this system of minute monopolies is seldom related to its cause. That effect is the disappearance of a public culture, and with it of the feelings of "belonging," of "community," of "togetherness," whose loss we deplore in these pseudo-technical terms. The very idea of The Public (as opposed to "the population") is growing dim, and as it fades one hears of such tragicomic predicaments as a publisher's not knowing for whom to prepare a revision of his famous encyclopedia. He must inquire, maybe take a poll: Shall it be for the specialists who write it? the laymen who need it? or the high school children for whom it is chiefly bought? The laymen no longer entertain a conception of general knowledge which the experts can assume or the children acquire. The logical end is for each man to talk to himself and hope he will understand.

Specialism destroys in still other ways. Lacking objects of contemplation upon which ideas and judgment are free to all, men can cherish no ideal of man, not even of the narrower ideal types —the soldier, the teacher, the jurist, the statesman, the divine. Each person is lost to view in his private crypt and complains of isolation. The cry is so common that it makes the "two cultures" hypothesis look like a utopian hope: if only all the scientists formed one clan and all the humanists another, the gregarious emotions might be so richly satisfied there would be no desire for a larger brotherhood.

But if the wish is to bring these two groups together, the scientists will have to forgo their exclusive preoccupation with research and discovery and be willing to work at the difficult task of making science an object of contemplation. They will have to begin the specialized work of organizing their ideas and finding a critical vocabulary for them—as has been done for literature, painting, music, and architecture; for ethics, religion, politics, and war; for history, for the social and private emotions, for games and sports and other pastimes, including cookery and murder. I see no signs today of any such transfiguration of science, none even of an awareness of the true need.

The obstacle to it is social as well as intellectual. I have heard more than one young scientist confess a desire to write about his work. Each was held back by fear of the profession. The man-eating clichés are ready to pounce: only the performer knows what he is talking about, because he is "inside"—which insures that he has never stood off and compared his work with other kinds.

It is true that no outsider, not even the closest student of a science or art, will have the *same* familiarity with its difficulties as the practitioner. The outsider is often wrong and sometimes unjust; for lack of perfect fluency in the speech of the guild, he can sound ridiculous even when he is right. But none of this makes him any the less necessary. Science itself would never have made head-way against the theological monopoly if the advocates of science had not gone "out of their field" to criticize philosophy and re-ligion as amateurs. It is the very familiarity with his own shop that prevents the professional from being critical of it or contemplative about it—let alone the fact that he may be an excellent practi-tioner without possessing the art and logic of public exposition. And this alone would make it an error to equate a possible cultiva-tion of science with the daily work of the scientific specialist.

The scientific culture in which we live, then, is one based on the particular brand of specialism derived through science from the older division of labor. The new excludes with contempt the public spectator and restricts everyone to his own expertise. So far as there is an ideal type of citizen, he is the maker, the performer, the know-it-all, who is inside and may not get out. Since it is obviously impossible to prevent people from looking over barriers, the ordinary man is aware of his peers' activities, but distantly and fitfully. He "follows" the news of discoveries and inventions, buys the advertised products of technology, and is buffeted be-tween their amazing results and their startling side-effects. He does not possess the bulk of the wisdom of his own people, but only a small part; what he can have is a multitude of objects bred by the collective knowledge.

Outside the United States this is called Americanization; but Americanization or westernization is merely the result of distributing cheap goods and installing machinery. The social upheaval that follows the machine changes many habits and some ideas and turns primitive or agrarian cultures into "scientific"; but it does not simultaneously turn old populations into scientists and technicians. Rather, what is loosely called materialism overtakes the society. This evil consists in an increasing dissatisfaction with what is, in proportion as the standard of living improves. Utopian expectations grow and make for a hard life; they stir up the resigned emotions and induce a perpetual futurism. This transforms hardships into "problems" for which solutions have to be sought. The obsessions of factuality, analysis, and abstraction develop. The statistical outlook prevails and establishes an equality that dwarfs the individual. In short, the Power and the Device seem to fail at the point where the life they are intended to sustain might begin to be worth living. The complexities of this unintended sequence are what we must now examine and illustrate.

III

Love and Hatred of the Machine

IN THE same vague way as our preachers decry "materialism," the layman decries "machinery." He feels hemmed in, pushed around, frustrated by it. Yet he is also proud of it, he enjoys and marvels at its products, and he reveres science, which he believes to be the source of these blessings. Himself a machinist, a technician, literally or in a figurative domestic way, he is of two minds about the miraculous machine that he and his ancestors have built and locked around their limbs. He loves and hates it at the same time and by turns. The first cause of these emotions is man's uncertainty about Work.

If shuttles and zithers moved by themselves, said Aristotle, there would be no need of slaves. Aristotle was not concerned about the slaves' lot; he was expressing man's conviction that to avoid work is imperative. Western man's stubborn effort to unload work upon machines argues a superior or Aristotelian gift of laziness. Yet by the same tradition man knows that work is necessary and that it fulfills the self. Work is virtuous and useful, and good works—not just good thoughts—are necessary to salvation.

At the same time it is praiseworthy to save labor by ingenuity; *ingenium, genius, engineer*—the words have had their present connotation since Tertullian in the third century. Nearly two thousand years later we still promise ourselves pleasure either way. There is delight, as our hobbies prove, in using a well-balanced tool and accomplishing a piece of work; and delight also in watching a great machine, intricate and smooth, seeing in it the symbol of Ingenious Man.* It is as if through machinery the mind's work

* The aesthetes of the nineties detested machine industry but admired the machine

31

were eternalized by effortless repetition.

But the inner struggle goes on: wielding the tool is not the act of the hand alone; the whole man works the ax or chisel, the scythe, the plow, the oar. The breath-borne rhythm, the concert of eye and sinew, the changing purpose and the sight of it fulfilled, the assured play of muscle that we find again in physical sports, are sensations good in themselves and they betoken art. Above all, the absorption of the self in manual work is unique. Alert yet only half-conscious, the craftsman knows bliss.

Compared with the humanity of handwork, which makes us one with the first woodcutter of the Stone Age, the stupid automatism of the machine is revolting. The machine's ceaselessness is as destructive of joy as its noise; its motions are a caricature of human gestures. Far from being self-absorbed when tending the blind frame, we turn into menials who must repress casual movement for fear of hurt. But again our feeling shifts, for we remember that manual labor is slavish and gross. Those condemned to it have least exemplified humanity. In every tongue the words boor, villain, navvy, clodhopper imply coarseness and connote the brutishness of work.

The earliest inventor, we may assume, was a cripple who, tired of being mocked by his thick-necked brothers, improvised the lever and the wheel: they would have heaved and dragged to all eternity.* His wits sharpened by shame and envy, he produced out of nothing, by leaps of original genius greater than any subsequent achievements, those fundamental machines—the pulley, the blade, the hammer, the needle, the spindle, the shuttle, the pump. Another originality led to the amenities—cookery, wine, and soap; also the comb and brush. And after the genius came the fiddling patience which perfected the tools and multiplied the amenities.

---

itself. Oscar Wilde, visiting Chicago in 1882, spoke of the beauty he found in the patterns and rhythms of machinery. Huysmans in *Against the Grain* asks whether any woman's form is more splendid and dazzling than the two locomotives on the Northern Railroad. Villiers de l'Isle-Adam, for all his satirical hatred of modern technology, praised the mechanical genius of Edison, who, as the hero of *L'Eve future*, manufactures a living woman.

* The soldier too invented, and war has ever been a great inciter to technology. But as a type the soldier remains an individualist, master of a nonrepetitive art.

And yet the taint of the original envious shame—the brand of Cain, the city-planner and worker in metal—clings to the whole apparatus. Some trace of guilt has always been felt: the devil's hand is behind the new process and secret formula; the tillers of the soil fear the weavers because they work indoors and use inhuman appliances.* Modern man is content to ignore his machines' secret; and when beyond childhood identification with Robinson Crusoe, he repudiates appliances and joins Thoreau at Walden. He calls it camping, hunting, roughing it—all aimed at freedom from our abject and obscenely interlinked contraptions: there is indignity in not doing certain things for oneself; we cannot love the cheap and fugitive gadget; we despise the ersatz masquerading as nature's stuff; we grow weary of the straight-cut, the exact machine finish that handwork proudly refuses. And in the recoil we misremember *Walden*. We imagine Thoreau cured of civilization and self-sufficient in the society of aquatic plants and a friendly loon. We forget that he took with him an ax, a sack of beans, and the clothes on his back, to say nothing of pen and ink. We forget the pleasure he took in hearing the telegraph wires of the nearby railroad humming in the wind, an unexpected zither singing by itself of modern industry.

Whatever the device, the task remains of feeding and clothing mankind. The original choice was simple: brutish manual labor or starvation. And so far, the only escape is powered machinery. Toiling and the machine produce different ills, but the end is the same, no more materialistic now than when Faustus sold his soul for food and clothing. He expressed the ancestral want, and till long after him no one thought of labor-saving machines as anything but philanthropic—not until the nineteenth century, when artificial power began its war on man. A special providence then seemed to inspire English clergymen in the making of new devices, but these inventors foresaw no more than Aristotle the immediate, horrible consequences—the destruction of rural life, the mass exploitation of the poor, the cancerous growth of cities, the

---

* George Eliot records this superstition as late as her own time. See the first chapter of *Silas Marner*.

uglification of the world through noise, fear, and filth.

It was soon noticed that the machine—the powered and coupled device—differs from the tool in that it gives abundance by taking away the sweetness of life. Robert Owen wrote in 1816: "The general diffusion of manufactures throughout a country generates a new character in its inhabitants; and as this character is formed upon a principle quite unfavorable to individual or general happiness, it would produce the most lamentable results unless the tendency be counteracted."[1] A new and degraded slavery annihilated the self. Endless pages of Parliamentary Reports record the physical facts; they are well known, and no less appalling for being familiar.*

The situation in which we stand a century later is not so clear. We are vehement about dehumanization, though the worst of the industrial evils have been corrected by law, by community feeling, and by the progress of technology itself. It has cleansed its soul— as well as its face and hands—thanks to new sources of power and the elaboration of machines and processes. But that very improvement means that industrial man does not in fact work; he does time; he stands in a pillory-with-a-purpose. The most sympathetic students of our modern technology are forced to report: "Man is not biologically or psychologically equipped to do such work."[2]

The machine thus deprives man at once of bodily exertion and fatigue and of the dramatic victory of work. Consequently man's feelings about himself during the greater part of his waking time are not joyful or single-minded. This is true not only of strictly mechanical work in the factory; it applies to all the tasks that aid

* "No. 116 Sarah Gooder, aged 8 years. 'I'm a trapper in the Gawber pit. It does not tire me, but I have to trap without a light and I'm scared. I go at four and sometimes half past three in the morning, and come out at five and half past. I never go to sleep. Sometimes I sing when I've light, but not in the dark; I dare not sing then. I don't like being in the pit.'

"No. 14 Isabella Read, 12 year-old coal bearer. 'I carry about one hundred weight and a quarter on my back; have to stoop much and creep through water, which is often up to the calves of my legs. When first down fell frequently asleep while waiting for coal, from heat and fatigue.'

"No. 72 Mary Barrett, aged 14. 'I have worked down in pit for five years . . . I work always without stockings or shoes or trousers; I wear nothing but my chemise; I have to go to the headings with the men; they are all naked there; I am got well used to that, and don't care now much about it . . . I cannot read or write.'" *Parliamentary Papers*, 1842, vol. XVII, Appendix I, 252, 439; Appendix II, 122.

and mirror industry. In them also men do not exactly know what it is they do under the name of work. Beside the machine, which is so exact and tireless and perfect, man grows weary and careless and indifferent. He cannot compete with it and has no incentive to do aught else. Production no longer depends on him. When he is away from the machine, its work and its output seem self-sufficient, yet demanding: the organization of its functions is a monstrous set of remote relationships. Substitutes for this angering drudgery must be found—real relations, contacts, conferences, talk.

For it is the principle of machinery, and not just its deadly chop-chop way of making goods, that is uncongenial to the mind and the instincts. A machine is a system and a breeder of system. It eliminates chance and, like Number itself, removes reality to replace it by abstraction. Not this artisan, but the work force; not customers, but The Market; not barter and trade, but "futures." The whole speculative, hypothetical character of modern life, its relentless fantasy, is implicit in the first machine.

No doubt one is oppressed as well by the size of the mechanical establishment surrounding us and by its incessant spawning of objects. But it is less the sight of too many humble ministers to our clay—food, blankets, shoes—than their abstract and imagined multitude that shames and nauseates us. The charge of material-ism mistakes the evil, which is not so much that we gorge as that we waste; industrial plenty does not make us sensual but jaded.* The love of luxury, the lust for fleshpots and broad acres and physical domination, has diminished rather than increased under machine industry. Despite the tireless multiplier that we command, our love of objects remains strangely mythical, ab-stract, immaterial; and in the most sensitive it is a source of malaise, not pleasure.

The gross aspect of a civilization successful at making things is of course conspicuous, and it has long been the main argument against modern societies. What the world denounces as American-ization now, the English were accused of foisting on Europe a

* Among the symptoms of this state of mind, one notes such publications as Norman M. Lobsenz' study of "The American Search for Pleasure" entitled *Is Anybody Happy?* (1962).

hundred years ago. Their harshness and crudity, their worship of money, their neglect of art, their superiority and complacency, were bywords. And as with us, it was at home that the system was most fiercely denounced. While Sydney Smith, the philanthropic clergyman, looked with some tolerance on "the porter-brewing, cotton-spinning, tallow-melting kingdom of Great Britain, bursting with opulence, and flying from poverty as the greatest of human evils,"[3] Carlyle, the unhappy mathematician-philosopher, lamented that "Our true Deity is mechanism. It has subdued external nature for us, and we think it will do all other things."[4] And Blake, with his eye on the children and the landscape, cried out against "The dark Satanic mills."

From the beginning, therefore, technological society was greeted with love and hate, despair and hope. Nor was the division only between opposing groups, but often within the poet, philosopher, or publicist. The poets particularly were captivated by the Promethean hope of human power overcoming Necessity, even as they protested against the spoiling of clear streams and quiet valleys. They could see beauty in the locomotive not as a brute force, but because knowledge and refinement were being carried by it over the globe, "perhaps to the displacement of war itself."[5]

Since it is impossible to separate the facts and the emotions, the machinery and the philosophy, I feel the need to replace the word "technology" by a shorter term, which will embrace the mixture of machine and soul at the base of our scientific culture and which will at the same time suggest our ambivalence toward it. For this set of conditions I propose to use the word *techne*, which is Greek for art, skill, regular method of making, craft, and cunning, as well as *a work* of skill, craft, and cunning. By using a fresh word, which has the advantage of not being a misnomer,* one can

---

* Technology properly means the theory or principles of techne, just as methodology is the theory or principles of method, and psychology the theory of the psyche. Hence *techne*. To use the "ologies" for both the thing and the discourse about it is part of that degradation of language I discuss later. Seeing this confusion, certain writers have used "technics" to mean what I would call techne. But apart from the involuntary association with *techniques*, the term "technics" suffers also from the parallel with such words as ethics and aesthetics, which again denote "the theory of."

refer to several related things in one breath—for example, the art or cunning of advertising which has developed with industry or "regular method of making." Again, when contrasted with science as the second great influence in a scientific culture, techne figures as "the art of science," that is, both the application of knowledge to needs and the foreshadowing of theory through mechanical ingenuity.

Before Techne could conquer, another idea charged with emotion had to become, first, congenial, then obvious—the idea of Change. We find it hard today to imagine the resistance to the inventions that befell mankind in the hundred years after Waterloo. Of them all, it was the railroad that radically altered the sensibility of western man. "Railroads, if they succeed, will give an unnatural impetus to society, destroy all the relations which exist between man and man, overthrow all mercantile regulations, and create, at the peril of life, all sorts of confusion and distress."[6] This editorial of 1835 was correct. It was the railroad which first coordinated space, time, matter, and men in the way we now consider normal and necessary: a timetable for every event, a name and number for every object, a common clock for all.

This new discipline was painful. A man born in the world of Squire Western, reared a farmer, suddenly found himself posted on a hilltop signal box, pulling levers, ringing bells, writing hours and code numbers in a ledger; "accepting" or sidetracking trains in order to make tons of coal and mail move at speed, while the human freight wove in and out in daylight or dark. Such a sentry-man serving twelve hours on end, giving hitherto unexampled attention to the minutiae of signs and symbols, was a prototype of the new man. Into his mind and sinews entered the headlong speed of the driven wheel.

True, older societies had understood violent coordinated effort in war; they had religious and civic ceremonies and trained performing groups to act and sing in concert. But the purpose was immediate and concentrated on one occasion. The performers knew and stood close to one another. Compared with this, the scattered army of railroad men obeying codes and signals was an abstract design, a dot-and-dash message compared with the spoken word.

As for the plain citizen, his molding into a mass man occurred in trains and terminals, where the masses first gathered and were subdued.* The quarrelsome humanity that we find entertaining in Smollett and Fielding makes us bemoan the passing of the individual, but its actuality would seem to us the anarchism of egotists and delinquents. Even about simple physical movement man had to learn some hard lessons. At the inauguration of the first English railroad, Huskisson, a director and a public figure, was run over by the Rocket: no one as yet knew how to gauge speed and distance, when and where to step. Man's reflexes failed him in the midst of powers stronger than the horse and the ox, and as devoid of sense perception as of animal friendliness. Never to be domesticated, the machine forced man to super-domesticate himself.†

The social and private ego died hard, dwindling with every device used for system and control. Thus in the forties that risky innovation, the ticket, replaced the human despatch notice written out for each stagecoach traveler. Today, when the IBM punched card has achieved the perfect abstraction of the individual—his life abridged as a series of little voids—modern man can see the distance he has covered along the path first marked out by the rails.

Huddled in trains and satanic mills, men found that their grimy work forced on them a new discipline of hygiene. Lacking the great outdoors, they devised the water closet and the bathtub. Herding together in factory and office removed the consciousness of sex and of class difference and prepared the raw material of democratic egalitarian societies. The boor, the apprentice, the artisan, the woodsman, disappeared in the undistinguishable, disciplined urbanite we find throughout the globe. The wordless calm exemplified in the ordinary working, by complete strangers of both sexes, of a self-service elevator in a tall building is the emblem of the long taming. Our self-control in public, our acquiescence in

---

* This promiscuity is what made Tolstoy liken the railroad to a brothel.

† The insouciance of natural man persisted for decades. The idea of safety enraged alike the passengers, enginemen, and directors of railroads. Taking precautions seemed unmanly, and even more important, it precluded impulsiveness and daring. The loss of these is one measure of our discontent.

social equality, our neutrality in the face of the other sex, did not come about by themselves. They are the manners of the age of mass transport, the social philosophy of the timetable.

Abundance, we know, was the original purpose and the inspiring vision of techne. And one consequence of success has been the curse of selling, in the new, absolute sense of the word. This necessity was seen quite early, by the first famous scientist to concern himself with technology, Sir John Herschel: "The advantages conferred by the augmentation of our physical resources through the medium of increased Knowledge and improved Art [Techne] have this peculiar and remarkable property—that they are in their nature diffusive, and cannot be enjoyed in an exclusive manner by a few. . . . To produce a state of things in which the physical advantages of civilised life can exist in a high degree, the stimulus of increasing comforts and conveniences and constantly elevated desires must have been felt by millions."[7]

To arouse desire, elevated or not, is the business of advertising, and it receives, with the machine, a hostility daily more violent. The continual irritation of our wants, the teasing mirage of Utopia, drives the most stolid of us wild. Everybody knows—and hates—the truth that advertising is inseparable from techne because mass production requires mass consumption, which in turn maintains employment. At this point too our culture transcends mechanics, since human desire appears as an indispensable fuel for the machine. That is why we are coaxed, shamed, hounded into recharging our flagging desires. We suffer twice, being goaded to buy and knowing the pains of pleonexia—the wanting always more.

Yet being, as we saw, ashamed of the surfeit of goods, we depend on advertising for the virtue and magic that objects lack as gross articles of consumption. Under Techne, what the people want from the fugitive products of their miraculous creation is a symbolic, ideal—I almost said religious—satisfaction. Advertising supplies the mythology of the creed. It tells the people that by consuming they will be lovable, beautiful, intelligent, gracious, learned, irresistible, confident, powerful, *right*. They will be blessed by the possession of certain objects, disgraced by the lack of them. The *charm* of the object—the literal spell—resides within it and one "wears" it like an amulet or a relic.

A successful piece of advertising, which Aldous Huxley has rightly judged to be more difficult to write than a good sonnet, supplies what is missing from the workless, sexless life desiccated by techne. It speaks of wonders and miracles in a world where science denies miracles and says the only wonder is that life exists in a cosmos not made for it. Advertising addresses an incantation—written, acted, sung—to the purchaser's trembling soul, tells him that the object was specially made for him, at the cost of extraordinary efforts and deep science.* The description of the thing, gravely technical like the directions for its use, expresses a concern, a tenderness for the user, which restores a little of the self-esteem that mass man can no longer feel for himself.

At the same time, the placarded appeals furnish an education through the eye, like the medieval picture book, which was the carved wall and stained glass of the Church. The manners, the uniformity of dress and of desire, the sophisticated platitudes, of the industrial population have come through the ad. Thanks to it both workman and tycoon take showers after work and probably use the same soap.

Taste and judgment naturally rebel against the mechanical flattery of advertising, the nonsense, and the transparent fraud; but the imposture begins with the product of the machine. Just as science renders experience abstract and to that extent false, so democratized goods supply an abstract, an appearance of satisfaction. All is for looks, color, façade. The bread is in a gay wrapper, all sliced, but it lacks wheat germ and tastes like blotter. Canned foods are luscious in the three-color plate, tasteless on the actual one;

---

* "Science rearranges wool molecules to give trousers a great new permanent crease." (Cloth manufacturer's poster in railroad stations, Autumn 1963.) "Your fountain pen is the result of as many as 255 separate operations involving close inspections and the assembly of many parts."

Perhaps the earliest, certainly the most typical, "scientific" advertisement was that which declared a certain soap 99 44/100% pure. The rhetoric of figures and the public's new awareness of what pure means in the laboratory permitted this arrogant modesty. Compare with this the contemporary failure of the name "Elijah's Manna" for a breakfast food now well known under a secular name.

girls beckon entrancingly from every billboard, wives are expected to resemble them only through detachable accessories. Sometimes the denuded object is partly restored: the green fruit has coloring added; the loaf is fortified with Vitamin D. The additive principle seems a new triumph of science. And as the old nobility kept the memory of chivalric success by coats of arms and mottoes, so every brand now carries slogans and insignia, the old and the new heraldry poeticizing equally dubious deeds. Those who see through the lies console themselves with the fable of Dostoevsky's Grand Inquisitor, who must go against his sense of truth that the people may have bread and miracles.*

The powers of techne having been everywhere acknowledged, the demands and the promises that join in the myths of advertising are frankly messianic. No one rebels against the perpetual futurism—"metals for tomorrow," "Century 21" (the theme of the Seattle Fair)—not even against the commencement speaker's forecast of "The Next Civilization." The mind is prepared by installment buying, which propels the customer into an endlessly acquisitive future—the realization of the pictured dream—based on unchanging life, health, and earning power, while insurance turns his present cash into future safety.

Techne is futurist because its machinery moves fast. Enough is made today for day after tomorrow, for next month and next year. The consumer must be found, held fast, and kept alive so that he can absorb the contents pouring out of cornucopia. Techne must not stand still but make and sell more and more, a frenzied addition miscalled growth. New products, new styles, timeliness, the sop to *Neugierigheit*, change, change, change—this is what keeps the demons whirling, whether they make shoes or publish an encyclopedia: no sooner has all of knowledge been dumped on your hearth in twenty volumes than it must be added to by immediate subscription for annual supplements. The reason is clear: to bring abundance cheaply the plant must be large; the invest-

* The people's wish to be fooled has grown more exacting in the last half century. Government control has improved, and medical knowledge is more widespread. The monstrous lies which a scholar has collected in an entertaining volume about nineteenth-century patent medicines (James Harvey Young, *The Toadstool Millionaires*, Princeton, 1961) are now fairly well suppressed.

ment is heavy; therefore everything must reach huge numbers; and because of all these magnitudes, life must be planned, which means mortgaging the future and living mentally in it.

However appropriate, this futurism is wearying. It splits the mind and makes the simple need for food, clothing, and shelter an obsession. Prudence is now a gigantic abstraction based on the statistics of group life and on the monetary form of time, which is credit. As a result, everybody plans; tries, that is, to manipulate the imaginary. If only something is as yet unrealized, people have confidence in it; it is the mood of revolutionists: *tomorrow* the state will be perfect. Today all we are sure of is that the age of the prophets has returned. Technicians, merchants, scientists, pugilists, politicians, lawyers fill the press with predictions and offer us a future graced with things no one wants but themselves.*

For each present and future advantage, western man finds that he must deal with a baffling Doppelgänger. For abundance, he must struggle with overpopulation. For the solace of change, he has to give up leisure, comfort, and quiet. For the excitements of futurism, he must endure the tyrannies of planning, youth worship, and compulsory obsolescence, including that of his own social self. For Utopia, he is put to the strain of always seeking the latest, the best. Altogether, he has lost the contentment that went with not expecting daily manna from heaven. And the loss is not the absence of a general quality of life; it occurs also as a set of positive ills. On this point our advanced literature has but one voice; its

---

* A sensible citizen recently said the final word on the subject:

"The proposal by planners to give us 200-mile-an-hour rail service on the East Coast is another of those romances that experts delight in, . . . but there is no need for another expensive foundation research project to begin railroad recovery. Any rider can provide expert advice, free of charge, to management.

"When it is cold, the cars should be heated; when it is warm, they should be cooled; where it says 'No Smoking' the conductor should enforce it; drunken and disorderly groups should be discharged; where it says 'Water' there should indeed be water; where there is scenery, it should be discernible through the window; where there is delay, it should be announced factually, as if to adults.

"Where there are sandwiches, there should be meat inside genuine bread, at a cost in reach of the average American income; where there are rest rooms, there should be sanitation somehow comparable to a well-run stockyard.

"When these simplicities have been faced, we may then call in experts, preferably acquainted with the better American and European lines, to suggest what to do next. . . ." John H. Cary. *New York Times*, Aug. 11, 1962.

leitmotif is that whatever is is wrong. Cultural criticism redoubles the complaint: "America is the nation that has supremely subordinated its institutions to the imperatives of technological innovation. Is it a destiny we actually want? Is it even a *human* destiny?"[8]

The predicament arises from the effect of large numbers. Too many human beings make each human being more indifferent to the rest, less pleased with his own life, less "human." This truth about numbers is not yet fully known or accurately reported. Henry Adams contrasted modern multiplicity with medieval unity and led us to suppose that the Middle Ages were on that account kinder to the human spirit. He should rather have contrasted medieval *multiplicity* with modern *multiplication*, which is its opposite.

Multiplication is unity over and over again, and its omnipresence today causes much of our pain. When we say "the mass" we mean numbers of indistinguishable people having the same wants at the same time and the same rights to satisfaction. This is the demand which denatures every act: one can talk and live with a few neighbors, exchange ideas in a company of friends, and attend to the feelings of a crowd, but a mass must be drilled and addressed like an army. Regimentation now haunts the fulfillment of every desire, the discharge of every duty. The use of literal numbers and of the machines that transpose the facets of life and of people into numbers is inevitable, and it increases the sense of living in a prison where the jailers are dumb, deaf, and blind.*

The first writers to give form to the modern rage against the machine were the German Expressionists, who saw around them the then most advanced, most disciplined scientific culture. They cried out against it

Nieder mit der Technik, nieder mit der Maschine![9]

They knew that the machine had escaped from its industrial cage and had taken possession of the street and the soul. But they did not foresee how the techniques they attacked would magnify the

---

* Consider as a symptom the prolonged outcry against all-number dialing and the new postal code called Zip. The latest of many protesters justly says: "Life is rendered intolerably complex by substituting number-memory for name association, which is not germane to the human psyche." *New York Times*, July 28, 1963.

horror by multiplying men. Since the 1920s, more than one billion human beings have been added to the earth's population. This "explosion" (we like mechanical analogies) is a direct result of techne—rising food supply, better hygiene, medicine saving the newborn and the middle-aged for long spans of life. These "benefits" swallow the hopes of abundance as fast as we approach it. We do not even know how to use what we have and in place of the old problem of evil, we face a new problem of good: the sought-for plenty rots or is dumped or is bought off at public expense before production. Abundance, in short, is here, but of no use except to devour its children. Pursuing Faust's original dream, we have lost our wager.

I speak of abundance devouring men, because techne implies a view of human beings which necessarily flouts the view that each person entertains of himself. Numbered, grouped without choice, stripped of differences not relevant to some temporary purpose, the individual under techne becomes in fact a manufactured product like any other, a creation of analysis and abstraction. And when the force of the multiplier crushes these fragments of men into a sort of pressed bale, the soul surrenders to the principle of techne itself; the mind accepts as fitting what I have elsewhere called "statistical living."*

At every turn statistics tell us the chances of death or disease, success or failure, accident, suicide, or crime to which we are exposed, not to say expected to conform.† When national strategy is discussed, we read in the expert literature of the "permissible destruction" of thirty-five to fifty-five per cent of the population by fallout and genetic blight. Well-trained, we rejoice that it is not thought permissible to wipe out seventy-five to ninety-five per cent. Men, as we say, are expendable; the next man is not anyone in particular, not even an enemy; he is "one of a thousand lifeless numbers," a unit to be turned into a cipher. Thus we are told by

---

* In *God's Country and Mine* (1954), Chapter 3.

† Notice how the critics of conformity fall into the very modes of thought they decry, for example, Mr. Raymond Williams who, speaking of literature, refers to "available methods of living," and "a lived culture." Adopting the operative view of techne, he calls the family "the community's agent in creating a desired social character in individuals." *The Long Revolution*, 1961, *passim*.

an authority on public health that the mottling of teeth to be expected from fluorides affects from 10 to 20 per cent of the users, which "is a rather insignificant public health problem." Or again, that the disease carried by pigeons causes only four deaths a year in Greater New York: out of eight million, a negligible loss. This bland indifference is linked with the industrial sense of the replaceability of parts and the ready supply of raw materials. How absurd in this frame of mind seems the old notion of each human life as a responsible pilgrimage.

Add to this machine-made helplessness the "trooping of emotion"* enforced by incessant mechanized communication, and you see how the technical conditions of abundance necessarily eliminate human variation and spontaneity. The improvement of method, the extension of system, the devising of standards, the measurement of performance, cannot help bringing all waywardness to heel.† The sum of these efforts oppresses even when it is impersonal. But though it is seldom remarked upon, the need and the opportunity to plan undoubtedly set free large amounts of aggressive emotion, of vindictiveness and the lust of lordly destruction—as Goethe showed in *Faust* when the cottage of the aged pair Philemon and Baucis falls under the improving hand of Faust the civil engineer.

The conflict within techne is thus the opposition of man to mankind. If we attribute to mankind the generous hope of abundance which inspired techne, we cannot help thinking that man is the principle of resistance to techne and its Utopian promise. And man in this sense does not mean the unhappy individual here or there who is bruised by the system, but man as the sentient Whole which is destroyed by the inherent needs of techne and large numbers.

The conflict between man and mankind is but a magnified reflection of our original yes-and-no feeling about machinery. The love-

---

* The phrase is James Russell Lowell's apropos of the electric telegraph. *Democracy and Other Addresses*, Boston, 1889, 24n.

† "Gray curtains . . . are issued with each apartment, and are required to maintain consistent appearance of exterior façade." Description of a private apartment house in Chicago, *New York Times Magazine*, Feb. 22, 1953.

and-hate has allowed the machine to escape from the workshop and invade what was once our native dwelling place, overturning our instincts under threat, like an army of occupation in our psyche. We are now "scientific" because the mechanical is the norm. Many struggle, but have no prop on which to lean. Helpless, childish anger is the only course left open, and this is what our satirists use. For them no longer gold but the machine is the root of all evil, the great soiler and depriver of life.*

The machine in any case forces us to unceasing self-defense. We use up life in response to signs, being artificially surrounded by matter in violent motion. Our sorry integument is never more than an inch away from things that burn and crush, cut and poison; we move among charged wires, adapt our speed to their commands, and remake our thoughts in the image of their broken idiom. We look for the light, heed the buzzer, turn the dial, distinguish color and symbols on lamps and cards, remember numbers and codes, decipher acronyms—no purpose is exempt: pleasure, travel, domesticity, the pursuit of art and learning. We perish, nullified, if we lose the serial number, the trade name, mislay the key or ticket or device or formula or token.

These are the requirements that overwork our nerves and make them an object of perpetual concern in common speech. Techne has relieved our muscles of strain at the expense of our nerve-ends. But our disorder is even greater than we admit, for in place of traditional work techne has made us pursue extraordinary forms of non-work. The primitive task of exerting force and mind upon some material object has disappeared in the abstraction of symbols on paper.† In work other than farming and mining—and sometimes even there—men are busied with signs; they neither toil nor think but watch and tend. Across the nation they form a collective nervous system for the muscles of the machine. What they give

---

* Thus Robert Osborn in the text of his book of caricatures, *The Vulgarians:* "America once had the CLARITY of a pioneer ax." New York, 1961, n.p.

† Except for passages in the writings of Peter Drucker and Theodore Caplow, little has been said about the sensations and emotions of work, especially the abstract kind I have tried to describe. As for factory work, sociological research has contributed the notable studies of Georges Friedmann. See *Le Travail en miettes*, Paris, 1956.

is alertness to signals in order that other signals may be transmitted in turn; everything depends on "the flow of information," on "data processing." We may see in this something like a projection of the railroad signalman in his tower, abstracted into a collective, impersonal anxiety and agitation in the dark.

Gradually all the professions, as distinct from trade and manufacture, are becoming one with them through the adoption of common methods and instruments. Facts and figures, reports, graphs, memoranda, pose the questions to resolve, which means: confer, confer, confer about the possible meanings of the multitudinous acts and wants thus recorded. For the machine system being an effort to satisfy a world aroused, it brings the single mind continually to grips with the many, from which it follows that the unconquerable enemy is confusion.

As the multitude generates what the executive calls pressure, so the failure to resolve confusions wraps him in the feeling he calls frustration—the will not served, nor yet balked by necessity, but embroiled in the machine. This pain, coupled with the sense of a great fatigue handed down by the nineteenth century, produces an inevitable and continual search for self-protection. In founding the factory system, as everybody knows, men wrecked the bodies of three generations of Europeans. What is less obvious is that they also used up a like amount of mankind's moral energy. The Victorians were indefatigable, as their plans, their talk, their writings, their *industry* show; we are fatigable in proportion. Shielding ourselves from non-work ranges from ordinary featherbedding to compulsive error, whose origin is anti-system, anti-machine. Since the size of most systems dwarfs anybody's contribution, it becomes negligible to him who makes it. Obviously, no one's defection will stop the machine, as no effort can finish its task.

In spite of these incompatibles, the men who are caught in "the organization" regard it as a fit object for their devotion. As we shall see, they speak of their labors as creative; they are jealous in behalf of the company and expect its employees and their families to sacrifice to it in testimonial ways. In return for this the organization gradually extends its benefits—not only pensions and profit-sharing, but medical and psychiatric treatment, further education, housing, sociability, entertainment—the full life around the empty heart.

True, even under techne there remains a privileged class of doers and thinkers—the scientists (including physicians) and the artists. They can have, if they choose, a fair chance of performing genuine work. But that work is more and more surrounded, cut down, by the incursion of "methods" and "pressures." The encroachment, which has already denatured the activity of teachers, lawyers, clergymen, civil servants, engineers, and prosperous farmers will not halt at the atelier or laboratory door. In greater or less degree everyone finds himself *busy living*, through identical acts-on-paper that devour the time that might have been leisure. Without leisure and without work, the being whom democracy professes to cherish, the person, is swallowed up in what is not even a crowd of people, but a set of functions. Whence the pathetic cry heard on every side, that this or that business or profession "is people," so clear is it that they have disappeared.†

Seeking an outlet, the desire for work changes into the "desire to help." Working "with people" holds out the hope of doing, each task having a beginning, middle, and end. Actually, the teacher or social worker or personnel manager is still dealing with numbers (that is, with too many); system is inevitable, and with it the usual frustration. The ever-present paradox is that with each improvement in system and machinery more and more people are employed to cope with others' troubles, neuroses, malfunctions. None of the ensuing relationships are precisely passionate, but they maintain the appearance of a community of men.

The workings of techne, which I must say again is not the machine as such, but the machine as we use it and think it, achieve the surprising result of a life in which intense activity is not distinguishable from dullness. It is not an exception but a confirmation that people still indulge the passion of the idler, whether rich or poor, which is gambling. Again, in this dearth of the joys of work one can find a cause of modern man's violent antipathy to power.

† See the program "People Need People," a broadcast discussion by convicts of their production of a play of that name, based on a psychiatric study. San Francisco Station KPFA, Program Folio for June 4-17, 1962, 6.

Political power, once denounced on liberal grounds, is now taboo in all forms, as if liberty needed no strength. The power of personality, of intellect, of age, of justly acquired position, is expected to be put down by the individual or the community like a mad dog. Due to fear or envy, the failure to come to terms with power leads to the unhappy shuffling we call insecurity, expressing by that name the lack of defined ego and the uneasy susceptibility to alien power. This insecurity forbids risk and roots out ambition, and this in turn bestows on social security—by which I mean both conformity and the welfare state—something like the unction of love requited.

One may wonder why modern man feels no vicarious power in the spectacle of his machines; that is, no pleasure in the reality of their power. The fact is that the sight affects him with the reverse sensation, not power but self-contempt. This is the feeling most widely shown, and even preached, in our literature, both cheap and fine. From literature it has passed into social life as a form of good manners. Nor should we be surprised that in so abstract a performance as techne the feelings of self-approval can find no target. Men will not prize themselves as consumers but only as makers. Some technicians can perhaps continue to enjoy the victories of ingenuity, which is a form of power. But it is at best the power of Vulcan the cripple, who hobbles on top of Olympus plotting devices.

Among his triumphs is a net in which he has caught Venus his wife and her lover Mars, and this is for the gods a good joke on all three. But for modern man this situation, too, holds nothing but pain. Like the crippled inventor, he despises his lot and is, moreover, trapped in his own net with the lover Mars and Venus the mother of Eros. The parable has historical support: machine industry, it was early noticed, has a peculiarly disturbing effect on the manifestations of love and sex. It is not easy to conjecture how an instinct and an emotion that are but slightly subject to reason can be affected by so abstract an influence as the machine. D. H. Lawrence believed that the sexual trouble came after the shift from the heart to the head that took place when industry devised its "rational" institutions. But no such shift occurred. Apart from calculation, inseparable from trade in all ages, the change has been rather toward the replacement of thought by automatism and routine, which should theoretically free the mind and let the feelings alone. It would therefore seem more plausible to suspect that the taxing of the nerves by signal and response, the steady hammering of the senses by the machine, the awareness of the multitude and the resulting abstraction of the self, began to alter the capacity, not solely for the act of love, but for it among other aesthetic experiences. Spontaneity deserted them alike.

Then, when self-despair and mental disorder came to be associated with sexual malfunction—the very phrase records our attitude—scientific self-consciousness set in. Statistics were taken, norms established, techniques recommended, and grim-minded performance became the partners' goal. Meanwhile the intention of birth control (now "family planning") kept the calculating spirit to the fore, while its practice exemplified the application of mechanics and chemistry to the sources of lust and life: love had progressed from the pastoral to the industrial stage.

Techne's remoter effects also helped to denature sexual love: the self-contempt of superfluous man, his voluntary repression of power, his lack of leisure and blindness to the quality of effort. After that, any shift of love to the head through the influence of psychiatry and literature was but the final step in a series that began with the inability to shake off the steady rhythm of the machine and regain, together with spontaneity, the natural sense of drama in work, the power to reach the top of the climax and exchange repeatedly, unhurriedly, achievement for desire.

Aware of this loss, the culture of techne supplied it by sexualizing the public through the enticement of the eye.* Techne is consistent in its habit of abstracting properties out of an article of consumption and then rejoining the fragments: as in the loaf of bread, take what is nourishing out of the flour to make it imperishable and then find "additives" to make the article technologically complete. The sexually neutral atmosphere of daily life is a con-

* On this subject, see Geoffrey Wagner's *Parade of Pleasure*, and on the role of abstract sex, the concurrence of two writers as different as Erich Kuby (*Rosemarie*, New York, 1960, 138) and Robertson Davies (*A Voice from the Attic*, New York; 1960, 283).

venience in the bisexual world of business, but it explains the need for substitutes, the joyless pornograms that range from advertising and movies to magazines, book covers, plain-wrapper mailings, and (for the elect) the quasi-clinical accounts of coition in the high literature of the past thirty years.

The double standard by which we perpetually judge the machine declares itself in still another contradiction. The strongest argument for techne has been that it would bring abundance; but second in authority has been the argument of Progress, by which the world has meant a continual change in the terms on which man wrests his living from nature. In older societies where this change has just begun, no doubt of its value is felt. In the October Day processions celebrating their liberty, the Chinese parade beautiful girls sitting in the angles of large machines.* In many parts of the western world the recent domestication of so-called labor-saving devices is still regarded as advantage without drawback. And the strong current of faith in the automatic improvement of all things continues to move the masses. They respond as before to what is new and also to the rhetoric based on the supposed impossibility of turning a clock back.† In other words, it is not generally seen that the only meaning of Progress is: keeping up the supply of conveniences to meet the ever-increasing difficulties of existence. Progress is measured—erroneously—from Then to Now, instead of relatively to the present predicament.

These naïve believers in story-book progress seem at times willfully blind. They "hail"—as they are the first to say—the sleep machine,‡ the irradiated codfish cakes, and the discovery that music

---

* My source is a film shown in Cambridge in May 1963 by Sir Gordon Sutherland, who had lately returned from a visit as scientific envoy to Peking.

† An English university student puts this thought-cliché in the appropriate jargon: "Modern Society is conurbanised man; a mass controlled by a specialized and centralized bureaucracy. . . . We *are* centralized; we are specialized; we are conurbanised; and the co-ordinating factor of all is Organization. Bewail the denial of Individuality . . . but modern society is geared to the speed of a mass-production line, not to the decentralized beauty of contemplating the infinite while making rush mats." Quoted in F. L. Lucas, *The Greatest Problem and Other Essays*, London, 1960, 323.

‡ Made by a French businessman: it is said to induce sleep in five minutes by means of an intermittent beam of light synchronized with the heartbeat.

stimulates growth in wheat. One ingenious mind regrets that no one has investigated the possibility of travel through the ground: "What we need is a mechanical mole." Another congratulates himself and us on rapid transport and a near future of total communication.* They do not see that the sleep machine is to recapture the fugitive, sleep, just as the teaching machine is a makeshift for the vanishing teacher. They do not see that the fastest vehicles jam themselves into immobility on city streets and that the one mail delivery a day, usually of letters that have taken two days to travel twenty miles, is retrogression compared with sixty years ago.

Similarly, the moralists who tell us that we have sold our souls for the "comforts and conveniences" that Sir John Herschel was the first to connect with science do not see that comfort and convenience are not synonyms but opposites. Techne supplies convenience after making comfort impossible. Comfort, as the old intensive phrase "creature comforts" implies, is a positive sensation and the relishing of it. A convenience is negative or *pis aller*. It helps us avoid trouble or supplies a makeshift. The paper cup is a convenience, not a comfort. Often an abstract idea of utility is all that one can find in a device, the use of which actually causes discomfort. We *think* we are kept warm, cool, well fed, or entertained. Modern conveniences have a social not an individual intention, and the intention excuses the performance.† A taxi company will use radio in cabs to round up as many passengers as possible, but those subjected to the steady meaningless quacking of the instrument pay for the convenience by discomfort. It is not then that one hears what an enthusiast of progress calls "the spiritual overtones of engineering."[10]

The unpleasant truth is that comfort is by nature a product of

* "It will be a world in which an individual, business, or government can establish contact with anyone anywhere, any time, by voice, sight or document, separately or in combination." David Sarnoff, *New York Times*, Aug. 8, 1962.

† In talk about television, its ability to delight the senses is always taken for granted. Critics dispute about the intellectual and artistic merits of the programs; commercials are attacked for their inanity. But few say that the commercials are offensive as interruptions, and that the color, size, and oscillation of any television image are incurably disagreeable. The human voice, moreover, is shorn (as in the movies) of the clarity and harmony that make it in life the exact interpreter of emotion.

manual work, of a domestic economy unjust to the many. Comfort is alien to techne, which, whatever other sins it commits, does not coddle the hedonism of the average sensual man. Anything but self-indulgent, we accept convenience as the essential condition of techne: there could be no mass distribution and no emancipation of the masses from drudgery without the innumerable conveniences that replace goods and services by partial substitutes. And Aristotle, if he returned, would discover that when shuttles and zithers move by themselves, something happens not only to the slaves, but also to the comfort of their masters.

Inured as we now are to these conveniences, from the car that we must pilot and pamper to the do-it-yourself kit that taxes our patience, we dare not take much pride in the system that brings them forth. Ours is not a technically high civilization. We have the knowledge but not the will, and shoddy manufacture and mis-representation perennially endanger the public. Drugs and pesti-cides have lately caused enough scandal for controls to be hurriedly applied. But the observant know that every object ushered into the market with the pomp and jargon of science should be pru-dently approached.† And prudence can only go so far: we are at the mercy of air pollution and adulterated food, of negligent de-sign and manufacture in vehicles and domestic machines. We take our chances, statistically encouraged to fatalism. This is a painful concession, for what is revolting about the badly made appliance is not solely that it is lethal, but that it violates the very art that brought it into being. We cannot help despising the pretense techne makes to rigor and precision, and we may well brood on the sacri-fice we consented to in order to reach a last state inferior to our first.

This is not to deny the special pleasure to be had from using a well-planned and finely machined artifact; but the experience is rare and expensive. The common sort grows less trustworthy as it becomes more "convenient" by being made automatic: the broken

---

† See the report of the discussion "The Government and the Consumer: Evolu-tion of Food and Drug Laws," American Historical Association, December 30, 1962; reproduced by the U.S. Department of Health, Education, and Welfare, June 1963.

air conditioner goes with windows that cannot be opened; the push-button car window that jams exposes the livid owner to soaking rain; the hydraulic landing gear must work infallibly or it dooms the plane passengers to dying in flames. So many examples exist around us of hostility to the human hand and human judgment, of disregard for human needs in the pursuit of an abstract idea of techne, that even the least impressionable consumer begins to feel affronted and alarmed.

The time is past, certainly, when Le Corbusier's famous phrase about a house being a machine for living would make the modern eye gleam with pioneering zeal.† A house, we have found out, is precisely not a machine, first, because a machine is not for living, and second, because living is what accomplishes itself without external aid. The recognition that this is so is sharpened rather than diminished by the kaleidoscope of blessings that our scientific culture offers. Its panaceas come—and go: sulfa drugs, antibiotics, psychiatry, love and permissiveness for children, DDT for pests, look-say reading, geriatrics, market research, "machinery" for peace, everything automatic. These offshoots of techne disappoint the hope of success that they arouse. Anger mounts, because the possibility we clearly see always outruns the deed, the deed which creates new impediments and new forms of distress. The airliner that mercifully takes us to the sickbed of the distant parent or lover or brings him medical aid or food, or rescues the shipwrecked, kills as many as it saves and fouls the air with noise. The news that is broadcast instantly over wide tracts only subjects everyone to the moral burden of the whole planet's woes.

To this physical and moral suffering is added the anguish of change itself, which makes more vivid than before the sense of time and mortality. For, we said, the counterpart of futurism is obsolescence, which in objects means waste and in human beings means a brusque removal from the main stream of life at what is now late middle age. It means also the tyranny of youth, necessarily

---

† Long before the architect's false dictum, an American engineer had spoken of a bridge as "a tool of transportation," and the prescient critic Montgomery Schuyler had warned against the effect that such a mode of speeech would have if it were transposed to a house. *American Architecture and Other Writings*, Cambridge, Mass., 1961, vol. 2, 349.

short-lived, but deemed as useful economically as the contrived obsolescence which it resembles. First and last, the machine claims its human sacrifices. Its onward march and our split emotions about it come together in futility. Indignation, scorn, satire, are of no effect, even when heightened, as they are in some of our writers, by the use of techne's aid—lysergic acid and tape recorders. Others express disaffection through yoga and Zen or a simplified, nudist, or highly sexualized existence. All these tactics aim at counteracting the literalism and rigidity of the machine, lest the mind give way before a palliative is found.* And the treatment in that event is only another application of techne: electric shock and tranquilizers. What Auden has called "The Chemical Life" is the machine-made remedy for Statistical Living.

We are not likely to reconquer the machine until we think and feel more intelligently about it than we do. A bare partisanship for or against, a sort of race prejudice in the matter, will not help us. When Faulkner took the occasion of an air crash to write to the *New York Times* an eloquent letter advising us, through the pilot's imagined last thought, "Let's get the hell out of here," and attributing the disaster to man's not daring to "flout and affront our cultural postulate of the infallibility of machines,"[11] the novelist was mistaking entrapment for subservience. Faulkner's view is that of the most articulate everywhere. But those who clamor for a return to the simple life commit the Fallacy of Utopian Addition: they join the lost blessings they discern in past times with the advantages we now enjoy and take for granted. Unfortunately, culture is one; the features of societies are not interchangeable. As modern Utopians we have a false idea that the past knew nothing yet possessed all the virtues. When Faulkner, again, interrupts a novel to denounce the modern American for caressing his car instead of his wife, one should ask who polished the Faulkners' carriages and curried their horses a century ago. And had he not also a wife?

---

* One is sometimes tempted to think that homosexuality is also a protest against the natural arrangement, which suggests too strongly a well-adapted machine and system.

When at times such comparisons are made they are apt to be full
of mirage, which confirms the impression that the protest against
machinery is itself an abstraction. Yet an historical perspective is
not hard to reach. When one sees death riding high at every cross-
ing, one should recall that the noble horse, through the agency of
the housefly reared on his dung, killed hundreds every summer
by infant diarrhea. When one is half-demented by the hiss of
compressed air and the screeching of brakes, one ought to remember
Boileau's agonized complaints of Paris noise in 1660 and Carlyle
and Schopenhauer's invective against the sound of whips and carts
and construction, well before the machine. When one repeats the
clichés about the City, as if agrarian life had been an endless pastoral, one should know that of the two million children still working illegally, most toil on farms. The city, loathsome as it is,
protects better against cruel exploitation, bestiality, and the flouting
of natural rights than the sparsely populated land.

As to the rights of the person, it is often assumed that in the
past everyone had a chance to exercise them, and thus that all mankind has lost dignity and stature by techne. This is to forget that
until recently the linking of human dignity with elementary conditions of physical well-being was scarcely ever made. The issue
therefore comes down to asking oneself whether one is ready, by rejecting techne, to sentence the world to the old inequality, starvation, and disease. If the abundance sought by western man is indeed
close and sharable, can one even in thought deny the rest of mankind its share? Can one say, I would destroy this power, these
devices, if I could, for the sake of a less troubled life? One can
of course renounce anything for oneself, pump by hand all the
water for one's needs and give up everything other than the vege-
tative life, but what sanction is there for making that choice for
others?

The moral thumbscrew which tightens here is that when one
reaches the point of asking such questions one has long partaken
of the goods that one proposes to snatch from the still-deprived.
To think of huddled poverty anywhere in the world and reflect
on the "materialism" that wages war on hunger, filth, and pesti-
lence makes one suddenly feel that these may be the remedy for

intellectual *Angst*. The feeling is mistaken, for misery may be reached by either route and it is not more imaginative to dismiss physical want than moral anguish. Where, then, does the true humanism reside? In the highly spiritual society of Tibet, as the brother of the Dalai Lama told the West not long ago, the energies of the leading men go into making thousands of prostrations before the Buddha and certain living saints. Wealth is not for almsgiving or artistic splendor; it is for the presentation of great mounds of expensive "good-luck scarves," and for burning butter in sacred lamps.

Against this extreme of anti-utility all western traditions, spiritual, political, and technical, have but one voice to say that the effort to pull humanity out of the mire must not slacken. The present difficulty is to survive the fatigue and anomie that have overtaken us as a result of our own heedless use of the machine.

Our attachment to science, we know, springs from our earlier attachment to machinery. In the modern mind the two are one. Their common features, the virtues they require, their joint relation to numbers and to trade, make of these originally separate institutions the single citadel of our culture. For all our anger against the prison which it also is, we cannot now choke off the impulse to diversify the skill of the hand through tools, magnify its power, and so command matter. The earliest triumphs in that kind were the greatest, as I believe, and as I am reminded every time I see in the museum the female screw-thread cut in the head of an ax by Stone Age man.

But the ax, despite Thoreau and Robert Osborn, could not be a stopping point. Pondering what primitive fire amounts to—a little flicker in danger of going out forever—the historian cannot but find miraculous the fusion of genius and will, of luck and cunning, in the searchlight beam that now pierces miles of hollow darkness. The other, the scientific, quest started with the desire to know the stars. To see in the dark, actual and metaphorical; to contemplate with more than a dumb wonder; to hold in the mind what the hand and the eye can seize between them—this is the task of science, which now requires the greatest of machines for its fulfillment. Thinking of this, are we entitled to say that our scientific culture, armed with the most prehensile techne ever known, makes up for the hurts it causes by giving to its children the pleasures for which Faust and the Chaldean shepherds strayed from the beaten path? This we can settle for ourselves only if we first consider how science today appears to the public and how it exists in fact.

IV

Science the Unknown

ALTHOUGH Techne leads to Science, the public's knowledge of the one and the other is haphazard and seldom deep. To use modern machines does not require an understanding of them, and the populace of the industrialized West is incapable of assembling, repairing, or describing the action of the commonest devices. Here, too, specialism is the rule; each mechanic is competent within his range and the public specializes in ignorance.*

This is not for lack of books, articles, and lectures on innumerable aspects of the sciences, most of them prepared with a particular audience in mind—children, the general public, the intellectual, the scientist or medical man outside his own field. This branch of publishing has greatly increased and improved during the past fifteen years. But what do the people know about the "revolutions" and discoveries of science once the article or book is put aside? What do they know of method and principles and what do they think of the whole enterprise? Even among one's scientist friends it would be hard to point to one or another who possessed something that could be defined as the standard or common knowledge of science among scientists. Medical men are perhaps the most likely to take an interest in science at large, or at least to be willing to talk about more than one. As for people who would reject the label of intellectual, but who none the less exchange ideas, they learn about science almost wholly from newspapers and magazines. They read the science column in *Time* or the staggering daily dose of science reports in the *New York Times*. They follow Lincoln Barnett's writings in *Life* or Rachel Carson's *New Yorker* articles on pesti-

* An eminent physiologist at a Swiss university once asked his class to answer three questions: Who was Spinoza? On what principle does a television set work? What teams are playing soccer here on Saturday? The percentages of correct answers were, respectively: 7, 20, and 98. Private communication.

cides and the balance of nature, having previously helped make her book on the sea a best-seller.

The people's interest, then, is discontinuous; it follows the news and responds to narrative. It does not come out of the science "taken" in school and college; it comes rather from concern about cancer, overpopulation, and bomb-testing; from amazement at manned satellites and transatlantic television; and from curiosity about the aspects of nature that "become news" when some feat of climbing or diving, or exploration takes place (Kon-Tiki, Mt. Everest, the Atlantic rift).

This reading public is naturally uncritical. It absorbs but does not assimilate. On Tuesday the headlines say that the universe is expanding and on Thursday that it is contracting. Overnight, the geological dates with which we passed the science requirement in college is shifted forty million years, doubling the length of certain past periods and halving others.[1] A little later the first ice age is found to have been not gradual, but sudden.[2] Unaccountably —to the layman—there is more time or space in the cosmos, less sediment in the sea. Again, Piltdown Man, on whom we had relied these many years for making a conversational point, is proved a forgery. And parity, the scientific emblem of mirror symmetry, is no longer an acknowledged feature of Being.

All this swirls around in our heads for a little time and finally goes down into the semi-oblivion where only the trace of a name is kept, for its next appearance in print. No doubt a good many technical terms get into circulation—Strontium 90, Carbon 14, ionosphere, plankton, DNA, hormone, allergy, carcinogen—but they serve common speech as mere pointers to whole realms of knowledge. Not one user outside the profession could give a fair definition.

Possibly the medical vocabulary is the most correctly known by the public, thanks to people's direct or vicarious experience. And medical research providing newspapers the best kind of scientific news becomes for the unreflective synonymous with science. The public does not distinguish types of science: physics and engineering appear as one, a confusion fostered by the cliché that Russia is ahead of (or behind) the United States in the race for "scientific mastery." And the engineering feat of launching missiles is taken

for one of the "conquests of science." Again, psychology and biology, sociology and market research, are all sciences, are all Science.
The figures that are uttered, the doctorate of the speaker at the
convention, certify it.

For by now the scientist is a well-characterized figure, as easily
recognizable as a postman. Yet, the scientist did not exist as a social type until well along in the nineteenth century.* The earlier
exemplars were universal men, men of affairs and of literature,
philosophers and poets combined. It was not more strange for John
Herschel to translate the *Iliad* than for Goethe to take up plant
morphology and optics. But by the end of Tyndall and Huxley's
time, specialization in science was becoming the rule, and soon
afterwards specialization in a subdivision of a single science. It was
then that the author of *Physics and Politics*, Walter Bagehot, gave
his memorable sketch of the scientist as he struck the common
imagination—aloof, thoughtful, above common ambitions, and in-
sensible to common feelings. What made a scientist, Bagehot
thought, was "a negative cause—the absence of an intense and
vivid nature. The inclination of mind which abstracts the attention
from that in which it can feel sympathy to that in which it cannot,
seems to arise from a want of sympathy. As a confirmation of this,
we see that even in the greatest case, scientific men have been calm
men. There is a coldness in their fame. We think of Euclid as of
fine ice; we admire Newton as we do the Peak of Teneriffe."[3]

Advertising supplied a picture to make vivid these subdued
qualities—the white coat and the absorbed gaze of the figure bowed
over a piece of apparatus. Most often he is a physician, the representative "scientist" of the time. For related reasons, one world-
wide hero of popular literature, Sherlock Holmes, helped perhaps
more than one suspects to fix in the common mind the features of
the scientist serving society while enacting the very epitome of
aloofness.†

* As late as 1904, Sir Clifford Allbutt writes: " 'Scientist' seems to me as
proper as 'artist' or 'naturalist' . . . but it should signify the professional worker;
hardly the great amateur, such as Boyle or Darwin." *Notes on the Composition
of Scientific Papers*, London, 1904, 3rd ed., 1925, 42.

† Holmes was not, of course, a scientist, but that is irrelevant. Compare Aurelius
Smith, Detective, whose "slender fingers dropped a slice of lemon into his cup
with the deliberate motion of science." T. R. M. Scott, *Bombay Duck*, New
York, 1927, 1.

Today something of the pose is gone, but only because it has been distributed, like other things, among a larger number of men, the multitude of experts. But even among the crowd the pure scientist remains for everyone a being apart: he is the man above reproach, more secure in that reputation than the artist, because science does not need temperament and is the reverse of erratic. Jean Rostand has told what "strange correspondence a biologist receives: people take him for a magician, a healer, a confessor, a friend. . . ." And he gives examples "at the same time comic and touching."[4] This is natural and even logical, since the public sees science and the scientist as exempt from the failings of mortality, as monopolizing the moral and intellectual virtues.

What in fact sets science apart in the minds of, I do not say the ignorant or stupid, but the intelligent laity, is the conviction that it embodies absolute exactness and rigor and is ever at one with itself. This must be so, because as long as debate continues no statement deserves the label that has come to be felt as stronger than "true": we say that a statement "is Science" or "is not Science." Science, moreover, is uniform throughout the world; therefore uniquely reliable and calm amid the foolish and tragic doubts of chanceful life. By both its method and its subject matter science is permanent. Under its gaze the transient vanishes, leaving the firm outline of a law or principle; science is the one foundation and rock bottom of all belief. In short, it yields true Knowledge, or, better said, the only Knowledge.

From these attributes of science, as it appears to the general public, a number of important popular beliefs follow. They are the ones that have struck the common imagination or that have received most attention from the popularizers. Thus there is the doctrine of the "three great hurts" inflicted by science on the vanity of man: Copernicus (or Newton) displaced man from the center of the universe, Darwin thrust him back into the animal kingdom, and Freud put him at the mercy of his unconscious.* These scientific blows are held to explain modern man's self-contempt and

* This cliché was solemnly intoned at the beginning of a recent film on Freud and his work.

justify his giving up will and responsibility: "scientifically speaking," he is altogether conditioned and helpless.

It is also true that periodically Heisenberg's "uncertainty principle" in physics is used to support exactly opposite conclusions, and in effect to give new validity to moral commandments and religious faith. Similarly, the sophisticated interpretation of natural laws as being an expression of statistical probability rather than of causal necessity, offers an escape from the prison of matter and motion. Every time that a competent scientist produces a book in which science is made compatible with religion and metaphysics he is sure of a large appreciative audience.* This shows how hard it is to say when and how much the public trusts the certainties of science and accepts its reduction of experience to particles and abstractions. But the inconsistencies are important elements in any description of our scientific culture.

Similarly, the cliché inherited from the nineteenth century, that science is but common sense refined and organized,† collides with such conceptions as that of anti-matter, or with the declaration that objects are not solid, but consist of whirling particles separated by wide-open spaces. True common sense abdicates without a struggle, and its disuse within its proper sphere leads, as we shall see, to socially dangerous acts and attitudes. The very dissemination of Galileo's discovery that objects fall at the same rate of speed (as if this were an obvious fact of experience which our ancestors were too stupid to see) is a symbol of the ignorance of science coupled with the loss of common sense. Let anyone throw a feather out the window, then jump after it, and he will see whether all objects fall at the uniform rate Galileo predicted. After which he may begin to perceive what conditions justify the scientific statement and how science deals with experience.

Disraeli was right, seeing the first effects of scientific publicity, to remark that the world had entered an age not of criticism but of "craving credulity."[5] When resistance to an overarching principle

---

* Cf. among others: Eddington, Lecomte du Nouy, Alexis Carrel, and Teilhard de Chardin.

† See, for example, Sir Humphry Davy's *Elements of Chemical Philosophy*, in which science is likened to both common sense and poetry.

is gone, there is no further criticism of the particular notions that seem to fit under it. Every statement in a form resembling the approved kind is accepted, and the distinction between fact, hypothesis, theory, guesswork, and nonsense disappears. Thus books professing to disclose the origin of life, the references to "the atoms within us" (who is the *us* that the atoms are in?), articles affirming with perfectly safe daring that "Science is Heresy" or, conversely, questioning: "Is Science to be Man's Servant or his Idol?" are received by our public without a tremor of doubt.[6] Both explanations and projects need only invoke or claim proximity with science to be taken on trust. Even intelligent readers think it self-evident that "music . . . is audible mathematics." And they believe thus, like the writer of the definition, because it is "no chance coincidence that nearly all the scientists I know are extremely musical in taste."[7] The casual correlation and the foolish metaphor do not disclose themselves to minds convinced that mathematics "being science in its ultimate form" underlies and somehow causes everything conceivable.

From these impressions to the confidence that science can do anything it desires is a short and plausible step. When alarmed at hearing about the menace of fallout, some at least are sure that science will make a pill to counteract the harm. And this provokes a physicist to lecture on what science cannot do.[8] But for one skeptic there are many, even among scientific practitioners, who do not hesitate to help debauch the public fancy. The same Jean Rostand who finds the ideas of his unknown correspondents "comic and touching" tells them that he hopes to influence favorably the functioning of the brain by means of appropriate substances: he supports Hermann J. Muller in wishing to change the character of human groups by the use of choice genes from great men; and he ratifies the prescriptions of Aldous Huxley's Brave New World, which he declares not a fantasy.[9]

The name and appearance of science obviously possess the authenticating power of the old theology, and the public may be forgiven for the extent of its faith. On the one hand, it is never taught, in school or later, to think about science; it has not even a smattering of the history of science, which might dispose of many of the confusions and superstitions now current.* And, on the other hand, the guideposts of common sense are either in abeyance or not applicable. Well-informed people like to think that there is a logic in things, a plausibility, which should guard them against error and quackery. But that plausibility vanished under the impact of techne's "marvels" and "wonders." The transmutation of elements, formerly a dream, is now a fact and a menace. And though we do not reason about it, everyday life leaves us defenseless against the improbable by making us familiar with the conversion of milk into cloth and spun glass into boat keels. There is no standard by which to doubt any well-turned hoax. When Mr. Orson Welles twenty-five years ago terrified the inhabitants of New Jersey by broadcasting the landing of Martians (without visas), he was only a little ahead of the crowd.† Indeed, the joke was more at his expense than his listeners', for in a scientific culture the cautious unbeliever will often be made to look foolish by later events.

But this does not change the great fact of our credulity, for even if life exists in other worlds and if subtle drugs and prime genes can make us worthier to live in this one, the present boastfulness of the expounders and the gullibility of the listeners alike violate that critical spirit which is supposedly the hallmark of science.

The enemies of mass culture may think that the ordinary citizen is no worse off for being credulous and self-contradictory. Why should he be a man of ideas?‡ But this contemptuous indifference

* For example, the persistent identification of Evolution with Natural Selection, as in the *New York Times* report of the death of Sir Charles Darwin, "the grandson of the originator of the theory of evolution." April 1, 1963.

† " 'Does life exist elsewhere and what is the form of that life?' We may expect [an answer] to begin coming in the lifetime of many human beings living today." *Life in Other Worlds*, a symposium sponsored by Joseph E. Seagram and Sons, Inc., New York, 1961, 3. Percival Lowell, one of the first to present astronomical observations in support of the view that life exists on Mars, did not live to hear the five scientists, including the President's scientific adviser, speculate about the notion for which he himself was ridiculed.

‡ Practical consequences follow just the same. Before the latest total eclipse of the sun, warnings were given out against looking directly at the darkened sun. Not only did some disregard the warning and suffer eye damage, but—more significant—many persons called the authorities at various planetariums to ask *whether it was safe to be outdoors during an eclipse. New York Times*, July 20, 1963.

does not extend to the educated, who are equally bewildered, if only because their habit of criticism can find no hold upon modern science. Their response to it, therefore, is passivity. They cannot choose but be credulous, though their superstitions are more complex and nagging.

Chief among them is the thought-cliché that whereas the physical sciences have attained near-perfection, the moral sciences lag behind. I shall discuss this fallacy in a later chapter, when I deal with its corollary, the effort to found a "science of man." That possibility is linked with a second, inconsistent with the first: the belief that "fact" and "value" are by their nature opposite and wholly divorced; fact being a natural product of the hard reality disclosed by science, whereas "value judgment" is "soft"—a free expression of our preferences. Presumably a truly scientific culture would not invent or condone the phrase *"Good* morning!"

But what is a fact? Let us hear the usual answer: "For scientific purposes a fact is anything that is potentially susceptible of measurement, and that is all that scientific work is about."[10] What this definition implies is that the mind can take the peas out of the pod of experience without regard to anything but their size. Yet to carve a something out of its environment and call it *a fact* is to make a judgment and to be doing it for a purpose which expresses value. It is a judgment again, charged with value, to note a relation that has not previously been shown or named. And since like other disciplines science aims at relevance, coherence, and simplicity of form, the facts and theories of science result from judgments of value. Fact and truth embody emotion as well as thought. This is what Pascal meant when he said that a proposition in geometry could become a sentiment. It is no doubt useful at times to separate fact and value, and it might be harmless if the distinction did not abet the tendency to make science the only knowledge, discrediting thereby "mere" choice, "mere" preference, "mere" value. The unspoken reasoning goes: "Fact is reality; the rest is illusion. Science delivers facts; *ergo,* other modes of thought deliver only illusion."

This habit, not to say reflex, of debunking is so fundamental in our society that it is necessary at this point to find out how it has become ingrained, having first shamed man out of believing the evidence of his senses. We call this "modern self-consciousness," but its working can be better accounted for by saying that the march of science has imposed on the mind an artificial concentration upon one sense, a specialization and indefinite extension of the sense of sight.

The new predominance of the eye, of the picture, over words and ideas, is perhaps the most striking difference in sensibility between our century and the past. We think and learn and talk in visual forms. The picture book and the illustrated paper are no longer concessions to young and unformed minds, but the pleasure of adult ones. Graphs and pictorial statistics accompany, when they do not replace, description. Graphs imply counting, which in turn means that the visible representation of facts classified and counted is now the normal ground of judgment and action. The intuition or imagination of reality is held to be "subjective," which is jargon for "unreliable," so that experience must be translated into numerical form if it is to be convincing. Whether the proposal is to open a shop or admit a student to college, figures somehow gathered are decisive.

And here we come upon some of the strongest reasons why the science that we do not understand is yet as deep in our minds as second nature. Techne prepared the western mind for science by making a commonplace of the machine analogy: the universe a machine, man a machine. But concurrently with this, the middle class, early promoters of techne, was developing a methodology of trade, which is a kind of precursor of science. It was the wool merchants who, it seems, inspired algebra by using plus and minus signs on their bales to mark deviation from a standard weight. It was the middle class which introduced the idea of making income match expense: it does not become an aristocrat to worry over such trifles; a true nobleman must live in a way that insures his being in debt— which is why the troubadours sang that it is shameful for a man to live within his means. Measure, number, weight, pennies, material goods, and no nonsense about debit and credit, inform the bourgeois idea of reality, so easily transferable to the laboratory and the patient and careful work of science. A fortune slowly amassed bit by bit, a cosmic scheme built up from small, neat, accurate contributions, suggest a likeness not accidental: double-entry bookkeeping embodies the rule of every equation just as the strict record of nature's transactions, large or small, yields a balanced account of energy and mass or any other related quantities within the system.

Nor is this all. Long before the twentieth century the expansion of trade had led to that curious union of matter and abstraction which is characteristic of our science. Money is an abstraction that stands for goods, credit stands for money, and their manipulation by complex instruments detached from goods or their use resembles the handling of scientific Matter: it too is no longer visible, yet it appears to the eye for close measurement through a multitude of indirect means. The device in each case is: convertibility.

Hence it is no paradox that, although we are continually called upon to act or understand by bringing the eye to witness, we live increasingly in a world of ready-made abstractions. What we are asked to judge is solid, concrete, substantial. What we are shown as proof is a graph or a model, a rate or an index relying for its truth on abstraction. Something visible is indeed before us, but it stands for something else that we cannot or need not see. The image of the little man in pictorial statistics does not stand for a man but for a hundred thousand men, and he is as "real" as before if he is cut in half to signify fifty thousand.

The ability to follow this shorthand of abstraction we owe to the products of industry and the rhetoric of their advertising. The lesson of percentages, averages, and graphic representation has been learned in the market place rather than in the school, and it has prepared us for that other scientific device which follows abstraction—making objects or agents out of what we abstract. Heredity and Environment are common examples; mutation, sex, the unconscious, the H-T index,* are the gods and demigods whose sport spins our fate. And the propensity to "reify" in this manner is not restricted to the physical and psychological. Whatever interests us in politics, in literature, in our inner life, we turn into a thing

* Humidity and Temperature, themselves abstractions. At first, the Weather Bureau published the figures as a Discomfort Index, but evidently came to see that discomfort is not a scientific abstraction. It is subjective, quite unreal, virtually nonexistent.

and speak of as an independent force acting or acted upon from behind the screen on which we see it. For this is a "picture" too, which helps us to come to grips with our difficulties and our questions. The book for businessmen asks: *What Makes an Executive?*; the magazine article wants to know "What Makes a Prima Donna?"[11] Not what *is* or what *does it mean to be*, but how has the product been put together, manufactured by man or nature? The supposition of an *unaccountable* prima donna, not made by anything we can analyze, does not occur, because she would then be *unassimilable*, being unique and to that extent unscientific.

Such a phenomenon would also be undemocratic. For the ethics of democracy joins naturally with those of science, trade, and techne in preferring its units equal and interchangeable. Only then can men and things be fairly counted for political action and social benefits. Statistical living is thus the necessary outcome of an industrial democracy under the authority of science, and not a perversity chargeable to bureaucracy as such. In the realm of mind, the trinity of trader, technician, and mathematician-scientist is the symbolic union of the powers from which our ways of thought derive. Regardless of any other intellectual or professional training, and certainly regardless of his scientific tastes or aptitudes, modern man is moved, quite as if he understood science, by its dominant signs and attitudes.

The remarkable achievements of science may well be said to offset the no less remarkable split in intellect that I am sketching. That is for each of us to decide, perhaps changeably in keeping with his mood. What must now be added to the description is the contrast between the scientific absolutism of common belief and the confused, uncertain actuality. Not only because pure progress has been held out to us, but also because we want it, we assume that science and machinery will always improve our lot. Only by an effort of imagination can we conceive the times when the unhurried march of mind concerned but a few and did not disturb anyone's existence. Being inured to change we encourage all the new "sciences" that promise to deal with our difficulties—which we invariably call problems, so confident are we that they have solutions.

Our strongest faith is that there is no limit to the good and the useful, no drawbacks to a proved success. By trying, trying scientifically, we shall end by reaching any goal we conceive.*

And when by luck or cunning we come within sight of our goal, we make a point of pretending that we have changed only what we designed to change, leaving the rest as it was. We refuse to count the cost, the errors, the unexpected and sometimes lethal complications. It almost seems a mean betrayal by nature when the miracle drug has unpleasant "side-effects." The phrase and its deferral to the end of the public report reveal our childlike concentration on the sweets.† The risks, that is, the disease and death, resulting from the use of pesticides, from the manufacture of nerve gas, from research with radioactive substances, from the injection of live vaccines, from blood transfusion, from sedatives like thalidomide, from contraception with blood-clotting pills,‡ from experimentation with drugs and serums in clinics and hospitals, are seldom brought into one view with the undoubted benefits, any more than moon-landing projects are measured in cost comparatively with other national purposes. All this is to be expected from a scientific-technical creed which is so much broader and more mysterious than science itself.

But this wild impetus which, in the eagerness to cure or earn the credit of cure, looses upon the public a thousand uncertain aids to life, also poses the more general question of the freedom of science—the right, that is, to study any part of life or nature by any likely means. Anyone who is convinced that science will in time master all subjects and answer all unknowns will readily grant science unlimited freedom. Yet reflection suggests that since science

---

* How frequently it is said, in our legislative halls and elsewhere, that if we could afford to spend as much on the study of cancer as in building armaments, we should shortly find a cure. That cancer research is lavishly endowed, that genius is not to be had on demand, and that a cure or preventive need not exist for every disease are considerations overlooked in our lay philosophy.

† "Heart Toll is Cut in Hormone Tests. Estrogen compound found to help men by altering levels of blood fat. 275 patients tested." And at the bottom of the column: —"Side-effects of the estrogen treatment include breast enlargement and decrease in libido and potency." *New York Times*, April 19, 1962.

‡ The *New York Times* reported this side effect on Aug. 4 and 7, 1962. It was denied on Aug. 9 by the American Medical Association. On October 23, 1963, the risk of cancer from oral contraceptives formed the subject of another report.

can offer no guarantee of success or safety, it may be desirable in the waiting period before Utopia to protect old-established interests, such as the right to refuse a treatment or to question the expenditure of public funds. Physicians object from the highest motives to the proposed law that would prevent them from making trials of new substances without telling the patient. Their argument is that the patient's knowing might interfere with both the control and the normal working of the drug. And without this unconsented risk of health or life medical research would not progress.* Both statements are doubtless true, but did the patient take a vow to aid medical research? Probably not. Society took it for him.

The question is not one of science but of civil rights. As such, it should be for the public to decide, on the basis of ethics and common sense, in spite of all the appeals to, and thunders from, scientific authority. But on such points the public is no longer clear, as we can see in the controversy that has been raging here and in England about putting fluorides in the drinking water at public expense. The object proposed is to help prevent tooth decay in children under ten. The rest of the population does not benefit; and there is dispute among experts as to whether it will be harmed. The important point is that the participants, including doctors, dentists, chemists, politicians, editorial writers, and simple citizens, have argued with passion what they all erroneously assume to be a scientific issue.†

In England, the Minister of Health has called the opponents of fluoridation cranks and fanatics; in this country, physicians who write on the subject to the newspapers fulminate against the unbelievers as if the Inquisition were back in our midst. To object to the plan is to be against science, that is to say, a heretic. Scientific fact is of course irrelevant to the issue, which is purely civic, and which

---

* It should be noted that once made public, an uncertain treatment cannot be thoroughly tested, since that would mean denying the treatment to a sizable control group of those exposed to the disease. The potent, and therefore dangerous, Pasteur vaccine for rabies is in that category.

† Having no young children to defend from either tooth decay or idiosyncratic poisoning, I am unconcerned about the outcome of this particular debate. Its bearing is more important than its contents, for it is clear that more and more political questions—for example, that of bomb shelters—will wear a deceptively scientific guise.

should be settled with the aid of quite simple questions, such as:

Is it common sense to treat by universal dosing a small anonymous part of the population, without knowing how much or how little of the treatment the intended beneficiaries will take?

Is it economically wise to put medicine in the water supply, most of which will be used to wash streets, flush toilets, and make beer?

And finally, is it right to subject everybody to a dosage of any kind without his consent? For there is no reason to stop at fluorides. The drinking water can also carry tranquilizers, laxatives, and aphrodisiacs, for the sake of giving chosen groups of the Children of Techne a happier life. One hopes that behind the fluoride scheme there are politics and selfish business interests: the presence of solid ulterior motives would restore one's faith in common intelligence.

The third of the questions in my list, What is right?, is as hard to answer in a scientific culture as in any other, and men of science who have recently come to feel the difficulty have remained perplexed. The charity patients who perish while making an unsolicited contribution to science constitute a moral burden, which is little lightened by the consolation "that's the price we pay for medical progress." For not only is the "we" ambiguous, as I remarked, but the acceptance of this justification means readiness to do evil that good may come, the end justifying the means.

Now, society cannot maintain itself without frequent resort to this principle, but it uses the state to apply it and it defines and controls the applications. Science, when it is free, is not so controlled. No one knows or can know how far individuals with great and relatively secret powers can be trusted, especially when they are high-minded and acting in a noble cause. The statistical view of human life and the devotion to science are capable of inspiring the most conscientious cruelty, as in Aztec sacrifices. Some observers indeed perceive a natural link between the instrumental use of animals, the delight of some surgeons in wartime ("the things you can try!"),* the spreading use of shock treatment (and its lay counter-

---

* Harvey Cushing (*From a Surgeon's Journal*, Boston, 1936) emphatically disputed the benefits to civilian practice of war surgery, which, he said, produced bad habits which it took years to overcome.

part, the "stress interview"), and the fact that a good many physiologists carried on live experiments on human beings in the German concentration camps.*

Long before that extreme is reached, all sorts of positive harm, ranging from the violation of privacy to the callous use of persons temporarily helpless, can be inflicted on a trusting population.† At any of these points the question of morality takes on the aspect of a philosophic one: is knowledge so desirable that it outranks all other goods? And that question in turn has economic implications. The thoughtful scientist and the layman whose taxes support him must sooner or later come to see that the social situation of science has changed since the nineteenth century. After winning its war of liberation, science declared itself answerable to nobody. Knowledge being good in itself, the searcher for it must give no thought to consequences. Scientists were uniquely privileged—they were in fact chartered—to disregard ordinary contingencies and restraints. The moral ones had in any case unpleasant associations with the creeds that science had battled with. So the slogan was "Science Absolute." Never again, perhaps, will scientists enjoy the aloof eminence of their state during the years 1870-1945: their one limit was that of means.

But after Hiroshima a series of grievous dilemmas confronted the scientist: "Should I work at an occupation whose fruit is *this*?" The choice lay between serving one's country and loosing destruction on the world. Next, the generous facilities put at science's disposal by governments forced a second choice, that between secrecy for national ends and the former openness for world enlightenment. Finally, science became the darling of industry, and here the scientist's duty to be secretive is to his company, which must compete in the market. When scientists are responsible to government, industry, public opinion, and private conscience, the old scientific freedom is manifestly gone.

---

* See a partial list in Sir Sydney Smith's *Mostly Murder* (London, 1959). As a postwar observer, Sir Sydney reports that the freezing and phosphorus burning, the inoculation with gangrene-producing bacteria, and other forms of utilitarian or sadistic research yielded "little of any value" (280, 281).

† Draftees have latterly become convenient guinea pigs. I recall meeting a friend of mine, a distinguished biochemist, who was returning from a mass inoculation at a training camp. He grinned as he described his satisfactory mission: "My God! were they sick!"

And though science has at last regularized its liaison with techne, the terms and offspring of the marriage fail to rejoice the reflective man of science, ill equipped to steer through the politics that now surround his natural interests. The cost of modern projects is so great in cash, space, and manpower that even wealthy governments must choose—shall we try for the moon in the name of Apollo or for the center of the earth under the apter name of Mohole? The public has begun to compare figures and raise questions, and for the first time a subsidized scientific station has been abandoned on grounds of expense.*[12] The result is that the scientist is catapulted into political competition and governmental responsibilities for which he is often ill-suited by training and temperament.† It is therefore an illusion to suppose that the end of scientific freedom, the union of science and techne, the awakening of scientists to social and moral issues, means the handing over to "science" (in its connotation of wisdom) of the troubles that have plagued mankind since the dawn of civilization.

In discussing the unsettled ethics of scientific work, many of my examples have necessarily come from medical research. The relation of doctor to patient makes the issue more vivid than that of the physicist to the unknown populations he may help to destroy. But there are critics of medical research who say that it is not a science, but an art which uses trial and error and simple counting for its success. Science is a loftier thing, unconcerned with the quick or the dead. The fact remains that to modern man the physician is science's nearest representative. What is more, to the ordinary citizen, to be a scientist means possessing absolute status in a body set apart. Its members are all deemed to "know science" and practice it with equal competence. There is no suspicion that as in other

---

* There is in the history of postwar research in the United States a famous mouse known as the quarter-million-dollar mouse, from the sum required to equip it for the disclosure of the knowledge it possessed.

† One of the first scientist members of the Atomic Energy Commission said on retiring that the hardest lesson of his first years of service was to make decisions before all the evidence was in. This scientific misbehavior required by worldly affairs was painful; but as he could readily see, to wait and do nothing is also to make a decision. He might have added that in the laboratory, but not in the world, one has an indefinite number of chances to do the right thing.

professions a wide range exists between a Newton or Faraday and a satisfactory but untalented research man in industry. Nor, again as in other professions, that the practice of science need not confer insight into its nature.* These superstitions are doubtless due to the conviction that it is difficult to be a scientist at all.

But in spite of its high place, the scientific profession does not constitute an elite, intellectual or other. The chances are that "the scientist," from the high school teacher of science to the head of a research institute, is a person of but average capacity. It is one of the great advantages of scientific method, as Bacon long ago pointed out, that it can raise ordinary ability above what might be expected of it: science is the democratic technique par excellence. It calls for virtues which can be learned—patience, thoroughness, accuracy. It tests what it does by conventional means—numbers and instruments; it guards against error by the communal sense of sight. The definitions of science, also a result of collective thought, are not supposed to be subject to individual variation and thus to ambiguity.† Once understood, they do not slip away. Finally, terms and techniques devised by one worker are ideally communicable to others. Hence with care and industry a man of normal endowments can be a satisfactory scientist. Vision, intuition, genius are not excluded but neither are they required. The average performer in science may never distinguish himself, or he may do so but once in his life; he is none the less useful and deserves the name of scientist. Intelligence and will power equal to his would make only a negligible artist or philosopher, useless not to say harmful, except perhaps as a teacher.

Even so, at a time when paper work is reducing most professions to one executive routine, the method and tradition of science maintain a notable difference: the scientist is a worker in the old sense of work, possessing as he does a manual discipline besides the mental. The scientist, moreover, is likely to be a hard worker. The laboratory is well named; those who labor there sustain one another's enthusiasm for the task, which demands continuity and long

* The public similarly thinks that capable instrumentists necessarily "know music," whereas it is possible and even likely that competent performance fosters a rather narrow idea of music in particular and art in general.
† But see below, p. 81.

hours. On the campus of any large university the science buildings may be known by the lights that shine late at night and on holidays.

Teachers and students of science also differ from others in having a place of work which tolerates few lapses and which readily shows up laziness. Similarly in the medical arts (in which I include dentistry), the manual work combines with the urgency of pain and danger to bring out in the practitioner industry and concentration. Doctors may not be busier or more harried than teachers or merchants, but as they go about filling a tooth, reading an X-ray, tending a patient, their efforts form the kind of toil that fulfills by proving its own use and success.

In being true work the activity of science is both above and below what the public thinks. When Dr. James D. Watson spoke at the Research Corporation dinner where he was being honored for his work on DNA,* many in the audience were startled by the account he gave of his career as a scientist. They expected a story of gradual discovery by orderly experiments and unfolding truths. They were prepared for setbacks of a technical kind which would retard but also heighten the final triumph. They were not prepared for the red thread of life which Dr. Watson wove into his narrative—the chance meetings and clashing personalities, the haste and envy, the fumbling and idling, in a word the politics and social vicissitudes of science.†

His was a performance worthy of a novelist and a salutary lesson for laymen and scientists alike. For scientists seldom take a three-dimensional view of themselves and never an historical view of science. They tend to think of their fate as pure, and they persuade themselves as they have persuaded the public that they are passionless men who deal with one another as they deal with their materials. They admit to failings, of course, and concede that the events imagined by C. P. Snow in *The Search* and *The Affair*

---

* April 18, 1962. He has since received the Nobel Prize.

† As to the rewards of "guessing and idling" Watson reported that at one point Dr. Linus Pauling sent a proposed solution of the molecule structure in the form of one kind of helix, which proved wrong. This error, plus a clear new X-ray from workers in London, spurred the Cambridge team to work out the consequences of a new helical structure, with the bases and hydrogen bonds definitely marked according to the Chargaff rules (which they had neglected), and in two weeks they had their "simple and satisfying answer."

are possible: jealousy, theft, and falsifying do occur but, they hasten to add, that is not Science. What science is, in their view, amounts to an earthly translation of the kingdom of heaven, where fitness and perfection rule and nothing is other than it seems.*

This ideal science has unfortunately been taken literally by many. They have seldom been told that actual, historic science shows the resistance to new ideas, persecution of innovators, and toleration of mediocrity that are the usual accompaniment of professionalism. Rather, they have heard eulogies of "science as a way of life," assertions of "the omnipotence of science," and similar impossibilities, such as that the scientist's discipline forbids him to believe on incomplete or second-hand evidence. Even if a hypothesis turns out to be right, as W. K. Clifford argued eighty years ago, it is wrong to have held it earlier, when the known facts did not warrant it. The history of scientific achievement gives the lie to this intellectual masochism, which is in fact unpracticable. It would make the teaching or learning of any subject sinful, since both require much to be taken on trust. Such an ethics of science is in effect a piece of showing off, which if followed would set the highest value not on truth or the intimations of truth, but on the process of demonstration. And since it is another (legitimate) boast of science that it is wise enough to alter its predecessors' findings, it is clear that the evidence can never have been all in. Scientists do therefore accept incomplete data and their conclusions are of this world.

Science and scientists themselves would benefit from not having to sustain a myth of perfection. The corporate blindness which delayed recognition of the work of Avogadro and Mendel, the persecution of Semmelweiss and Raspail, the reversals of opinion on cosmic or medical questions, which now have to be shrugged off or concealed, would then be seen as errors of passion and intellect resembling those that lead to the neglect and harassment of many

---

* Scientific discoveries may on the contrary be at the outset a source of new error, as for instance in the conviction of innocent people as poisoners when the Marsh test for arsenic was first used and the natural presence of arsenic in many household articles was not known; or when the shadow of Pasteur's authority made Scott and his fellow explorers boil lime juice and thus destroy the vitamins that would have saved them from death by scurvy.

great artists—a fact of cultural life, not a flaw requiring anyone to say that any failing of science is "not Science."*

As for educated opinion, it too would profit from a sober knowledge of the limitations of science. That knowledge would not lower, but heighten admiration by supplying a correct view of the difficulties overcome. The merely approximate accuracy of all scientific determinations, the fact that investigators can rarely obtain the same results from the same experiment, the recognition that convenience, accident, plausibility, and "subjective" intuition contribute to the happy results, would ease the strain of uncomprehending awe at the thought of scientific method, at the very words "experiment," "determination," and "results"; and would relieve the lay mind of the unwarranted humility which it feels—or sometimes puts on to cover envy and distrust.

Nor would it injure science, but rather strengthen it like any other intellectual activity, if the habits, attitudes, and implications of science took their place as topics of public discussion. This is not to suggest that the technical part of the work be subjected to ignorant criticism, but that public issues arising from scientific research and findings be recognized and treated as such. At present poets, playwrights, novelists, historians, learned judges, philosophers, and divines are properly at the mercy of anyone who can follow what they say and is moved to comment upon it. The work of these professionals is no less technical, no less patient and careful, than that of the scientists; like theirs it receives technical criticism which few outside the guild can appreciate. But in addition the bearing of the work is for anyone to discuss. To except science from criticism is either to give it the status of an established religion, which no one is advocating; or to deny it that intimate connection with our lives and thoughts which the best scientists plead for, and which would make our culture scientific in a more complimentary sense.

---

* A cultural fact of the present moment is the case of Dr. Velikovsky. The merits of the scientific issue do not alter the deplorable treatment that his ideas received from the profession. The pre-judging, scorn, and closing of ranks against the heretic showed typical guild animus—as happened ninety years ago to Samuel Butler. But we progress, for Butler's sympathizers among scientists and publishers were not forced to resign their posts. For the history of the current case, see *The American Behavioral Scientist*, September 1963.

Science is either a great institution with which we should grapple, or it is a tribal creed that rules us through our ignorance.

The observer today is unfortunately hard put to it to find a boundary to that ignorance. Almost everything he hears or reads exemplifies it, and much of it bears the approval, tacit or express, of men of science themselves. Contradictions, absurdities, empty rhetoric possess the mind and come out as superstitions. Who are we to pity the hopes and terrors bred by a false cosmology in the Middle Ages, when we are daily reminded, on the one hand, that Newton, Darwin, and Freud have stripped man of his will, purpose, and dignity, and on the other, that this creature which is no more than a machine is about to enjoy a millennial existence under an endlessly improving Techne? The "revolutions in thought" ascribed to these three (each of whom, incidentally, was a scientist in a different sense) have not taken effect directly upon the people. Assumptions and inferences have seeped in through other channels than conscious ideas. Technical terms, metaphors, and stereotypes have done the work, all uncritically received after the first shock to convention.

The same uncritical passiveness now favors the currency of the cliché about science as a way of life. What it proposes, in answer to a fresh curiosity about science, is that a man of sufficient mind and will power can order his life according to the method of science, and hence that all men could do it. This is a prime example of the evil of metaphors in an age bemused by the scientific. No scientist could survive half an hour outside his laboratory if he tried to apply his habitual tests to his common experiences—analyzing, measuring, questioning such things as his neighbor's truthfulness, his tradesmen's honesty, his wife's fidelity. A man's relations with his family and friends would come to an abrupt end, his private pleasures would be destroyed, if he were even for a moment scientific about them.* It is even doubtful whether his health and temper could continue unimpaired without the aid of anti-scientific attitudes—the beneficent errors and illusions, petty prejudices and amiable over-

---

* One might have told the Atomic Energy Commissioner that he had made "political" decisions before going to Washington—when he married, bought a house, or accepted a post before the full evidence was in.

sights, that permit life to go on. Science is no more a way of life than stamp-collecting or the craft of embalming: nothing less than the miscellaneousness of life can be a way of life. And of all the trades and professions, science, as the most abstract and artificial, is the least fitted to supply a program for existence.

To some, it may be, "way of life" does not mean what it says, but rather the omnipotence of science. These men, including scientists, think it inevitable that in time science will rule every department of life. This prediction dates back a century. In 1848, the young Renan, freshly escaped from his seminary, foretold in *The Future of Science* the replacement of all the arts and crafts by the certified knowledge of the new masters of reality. But Renan did not think the time ripe to publish, and he withheld his book until 1890, when much of it seemed platitudinous. Two decades after Renan, Haeckel was still confident: "The number of world-riddles has been continually diminishing in the course of the nineteenth century through the aforesaid progress of a true knowledge of nature. Only one comprehensive riddle of the universe now remains—the problem of substance."[13] Half a century beyond this summit of faith the perspective is blurred. Despite the persistence of a certain rhetoric cultivated by business and philanthropy,* there is less of that "self-confidence among the learned" which, according to Whitehead, is "the comic tragedy of civilization."[14]

But there is still nothing like general recognition of the variability and impermanence of science. Only specialized students know how the scientists of the fifteenth century neglected and misrepresented those of the fourteenth, only to be misconceived in turn by those of the seventeenth and others down to the present. The assurance still prevails that we stand at the top of a regularly ascending edifice, even though, as René Dubos has several times pointed out, in ten years most scientific articles prove worthless, not because they are surpassed, but because they employ local techniques and use faddish terms and ideas. Mathematicians concur, and a famous

---

* See Warren Weaver, *A Great Age for Science* (New York, 1960), in which it is said that "we are just in the process of gaining a scientific picture of the total ascent of life," and "seeking to determine whether there are other and possibly more advanced beings" in "inconceivably distant parts of the cosmos . . . trying to communicate with earth-bound men."

chemist has urged that these erstwhile contributions be "incin-erated." Change is the law of life, and the whirligig of opinion in science follows suit, however sedately: biologists note that the rela-tive weight given to heredity and to environment differs every few years, and that the meaning of terms is not always constant from one worker to another; physicists vainly deplore fashions in their sciences, unable to explain why for the past twenty years it is quantum theory and not relativity that has been "the thing to do."* Whatever the reason, it is no longer permissible to rest the weary mind in the fixity of science.

To this further inherent cause of general ignorance is added yet one more. Specialism, as everybody keeps telling everybody else, prevents anyone from knowing all that "science" knows. What is not so familiar is that even if the best scientific knowledge were shorn of detail, it is anything but certain that the parts would fit together and approximate a whole. The parts are an avalanche of reports from a multitude of viewpoints, which at present nobody is concerned to sift and construe. According to the latest inventory, which is already a dozen years out of date, there are in the world 50,000 scientific journals publishing some two million articles a year—forty thousand a week. To cope with this production two hundred organizations make abstracts, the single volumes of which often equal in size an unabridged dictionary.

In any one science the purport of these abstracts themselves is far from known by any single mind, acknowledged master of the science though he may be. The upshot is clear and it cannot be bet-ter put than it was by the famous biochemist Erwin Chargaff: "A unified and consistent vision of Nature has become impossible in our days, at any rate for working scientists. Ironically enough, the only universal scientists left are the publishers of scientific books or the writers of science fiction. Each science protects itself from its neighbors by a cordon of slogans and catchwords; and fashion dictates whether this year we are featuring enzymes or proteins or

---

* Accordingly, when C. P. Snow wishes humanists to "know the meaning of mass," one wants to ask which meaning he hopes to have understood—inertial mass, energic mass, or good old-fashioned nineteenth-century mass? A large and learned work has been written on the meaning of mass alone. Max Jammer, *Concepts of Mass in Classical and Modern Physics*, Cambridge, Mass., 1961.

nucleic acids and whether we wear the molecules long or short. New journals are born every day by Caesarean section performed by skilful publishers; and as new disciplines are formed, so are new and mutually unintelligible languages: a Tower of Babel made of paper."[15]

To this glimpse of the scientific workshop in mass production must be added the fact that the purest and clearest of the sciences, mathematics, which had hitherto kept its coherence and unity, has begun to experience the same confusion. Disquiet about the expanding chaos has been expressed more than once, but no scientist has yet set aside his research to ponder a remedy. By the present code of the profession it would be tantamount to deserting under fire. For the old faith of the first seekers still holds, that the separate small bricks of knowledge will one day raise themselves into a completed edifice of truth. "I believe," said the distinguished physicist Mme. Wu, on receiving an award, "that these [discoveries] are but stepping stones leading to a deeper . . . understanding of nature. When we arrive at this understanding, we shall marvel how neatly all the elementary particles fit into the Great Scheme."[16]

The Great Scheme, for the present, is only a phrase. It no longer exists as a vivifying prospect in the mind of either the lay observer or the experimenter. Indeed, as a number of scientists have been heard to say in moments of fatigue, the researcher is as discouraged by discovery as by failure: success only opens a door into a vast room with another door at the farther end and so on in endless receding. "The greater the circle of understanding becomes," Dr. Chargaff remarks, "the greater is the circumference of surrounding ignorance."[17] And he adds: "But as our sciences expand, their specific gravity appears to become lower"; that is, they decrease in weight and importance relatively to the goal of organized knowledge. It may be that man's passion for knowing is fated to be baffled by an abundance of science as his desire for material ease is being baffled by the abundance of manufactured goods. If men survive the second abundance they may turn away from the first to some other way of assuaging curiosity. Science will then appear as only one historical form of *libido sciendi*, the lust for knowing.

But speculation about the mind of posterity will not tell us what we most want to know about the heart of our scientific culture, which is: the true nature of science. What is it at its purest, what varieties of human traits respond to its lure and modify the human urge to inquire, what nuances of thought and feeling attach to its elevated satisfactions? What consequences follow for man and society from its headship of culture? To appreciate these in full one would have to be a great scientist and an introspective one. Failing that distinction, one must be a sympathetic observer and listener, who for the moment forgets the external relations of science and is willing to look at the ideal discipline as it exists within the temple and the sacred grove.

V

Science the Glorious
Entertainment

THE affinity of science with machinery and with trade
has been touched upon and is not hard to see; its likeness to art is
more hidden, yet not less often claimed, especially by men of science.
It remains vague and honorific. The actual resemblance between
science and art lies in this, that the creation and purport of each
are full of mysteries. Some men of science, and also some artists,
think that they fully understand what they do; but others
doubt this and debate continues. Does science grapple with reality
or only with its own constructs, some of which are then found to
match reality by a sort of recurring miracle? Is science based on
a few assumptions, or does it repeatedly *presume* what the universe
is like, regardless of what may be? Do the elementary particles
exist or are they merely "features of a conceptual scheme"? What
is a law of nature? This last question is a kind of summary of the
previous ones, which are variants of a single doubt as to the meaning of science. It leads to another, What is scientific method? the
answers to which remind one of what Pascal said of poetic beauty:
on ne sait pas ce que c'est. The usual way of speaking about science
implies that there is one method; the record of discoveries suggests
that there are many methods, linked with differences of genius, as
styles are in art.

Contrary to common belief, questions about science do not receive
single answers any more than do the metaphysical questions that
science is thought to have taken over from philosophy: Is the
universe all matter, which now means all energy? How did the
cosmos come into being? Is it infinite or finite? Is energy "dying

out" as the universe expands, or is "creation" continuously repairing the loss? If loss occurs, how is it conceived in an enclosed
all-material system?

Scientists frequently express opinions on these matters, but most
investigators prefer to say that pure science has nothing to do with
ultimate philosophy. The purpose of science is to ascertain facts
and its method is to determine relations and express them in equa-
tions. The scientific sentiments are neutrality and impersonality, or
in one word: objectivity. Science solves the problems it sets itself
by leaving the observer wholly outside the experiment and submitting its results to all observers. To the extent that there is a
method, it is that of the "open secret"—each man closeted with
nature but all invited to look.

For science, as I said earlier, the eye is indispensable. Man is the
eye par excellence; animals, mostly color-blind, rely on smell, and
the senses of plants are a form of touch. From his other senses man
infers the presence or subtle properties of things; from sight he
thinks he has direct knowledge of reality. Sight is the controlling
sense, that which makes us *attend*; we have to shut our eyes to
sleep. We cannot turn off our other senses; but in sleep they attend
feebly and let us dream. Above all, sight is the exact sense, the only
one which permits minute comparison: that we could take refined
measurements without the aid of sight, or magnify the sense of
touch, is hard to believe. Equally important, sight is the sense for
ready agreement; the eye is "the best witness," as Dryden said,
which is why idolatry is a congenial error. And lastly, through the
common experience of infancy which connects sight and touch for
self-protection, we associate whatever we see with something hard
and permanent "out there." We reach out accurately to touch it
and it subsists in the darkness behind our backs.

From the mind's dialogue with the tangible-visible the scientific
outlook developed, juggling at first with the objects of common
sense.* But first or last, science clings to phenomena, that is, to

* A good example of common-sense science is the pleasing series of experiments
by which Dr. Charles Wells sought the explanation of dew. He stretched a sheet
a foot or two above the grass on a clear night and found dew on all sides of, but
not under, the sheet. Presuming that the air is as moist above as below the sheet
and its temperature is uniform, he inferred that one of the conditions of the

appearances, and finds or makes objects of what it sees. The word "theory" means seeing; theorem, a sight, an object of study— both words related to theater. From being watched, *theoria* came to mean a procession; and it is fit that scientific theory should now designate the vision of a process. Before he achieves theory in the correct sense of the term, the investigator must test the regularity of the phenomenon that interests him and must try to relate it to others by supposing a fixed sequence or connection.† In fortunate instances the link can be expressed in numbers obtained by measuring the various parts of the system, their relations in space and time.

The assumption about the world so studied is that it is a machine composed of parts in regular working order. The parts are all of one substance or matter, and all their doings can be measured provided the inquirer is ingenious enough to detect what is a part and what are its links. No influences of a different kind intrude, and certainly none of a fitful invisible-intangible kind. Matter in motion and no nonsense besides. Which is why all would-be scientific experiments in telepathy or extra-sensory perception are doomed to failure from the start: these things may well exist, but their conditions do not meet those set by science. Many observers say, for example, that for telepathy to succeed, the receiving subject must be in a favorable mood and must not be a skeptic about the enterprise. To say this is to deny any mechanism of the scientific sort: mood is not matter and skepticism has no wave-length.

Classical science, then, excludes anything emanating from the observer. He does not exist. Rather, his whole being is made into an anonymous eye, directed by mind, to be sure, but devoid of individual characteristics. Scientific method is essentially democratic and leveling. Hence it has often attracted men who were not given

---

formation of dew must be the open sky. This tallies with the observation that dew will not form on cloudy nights. Since these conclusions of 1814, refinement has been brought to the study of this diffusion of moisture, and measurement has been attempted, but the observation and reasoning have not changed.

† Theory in science is properly a completed network of facts and general statements about them. Popular usage makes theory mean the supposition that comes earlier, for which the right word is "hypothesis"—what is tentatively "put under," "sub-posed."

to introspection or strong feeling.* The very desire for simplicity and clarity and fixity often implies a certain reluctance to enter the hurly-burly of human affairs. And for those with a more versatile nature, devotion to science has been a Puritanism of the intellect, by which I mean the effort to repress the will-full emotions —hope, fear, haste, ambition, and self-deluding joys. For mankind at large this is indeed the great triumph of science—the unexpected miracle—that through some gifted and dedicated men, Man should make himself a disembodied eye before the universe. How unlikely a development of the restless rampaging beast!

For here is a different humility from the grief-stricken, breast-beating sort, in which ego so persistently lurks. In science, man's ego is put into the method and as it were socialized. The discoverer barely discloses his modest presence, murmuring only that "the facts speak for themselves." He is sure that he takes no part in what happens; he merely provokes an interplay of natural elements, and he stands ready to note results regardless of expectations. Thus his fellows are not required to believe in *him*, nor are they asked to share an intuition difficult to communicate. They can see his findings, caught in the tight mesh of figures. No wonder that this form of knowledge has come to be looked upon as the only one worthy of trust.

Free of the burden of persuading, science attracts the curious and contemplative minds who are eager to do their best quietly, and without competition or compromise. Such a detachment from the world, which yet serves the world because it is no escape into dreams, must appeal to any man of imagination. To be busy finding out the shape of a genetic molecule, or to surmise that there is an unknown mountain range in Antarctica or a shell of dust around the earth—to press against these doors opening on the unknown is to breathe an atmosphere of innocence and to know a childlike bliss.

---

* Bagehot's description is very just: "A tendency to devote the mind to trees and stones as much as, or in preference to, men and women, appears to imply that the intellectual qualities . . . which can be applied equally to trees and to men, to stones and to women, predominate over the more special qualities solely applicable to our own race—the keen love, the eager admiration, the lasting hatred, the lust of rule, which fastens men's interests on people and to people" (1856). *Works*, ed. F. Morgan, II, 59.

But although the scientific method is in one sense a prolongation of infantile trying out, its mature objects are not so easily come by. Few experiments are as simple as Dr. Wells's with the dew, and even the ideas of simple experiments—such as Galileo's with the inclined plane—are products of an uncommon subtlety. They depend on the power, first, to select from experience what is amenable to the method, and next, to abstract from this slice the operative unit or descriptive rule. By abstracting, the scientist disregards differences in order to collect similars. For the "mechanism" or "system" which the scientist wants to bring to light is hidden within the phenomena, and his talent consists in making it visible by such an alteration in the phenomena as will yield measurable similars. For this purpose, he takes small portions of the visible and applies to them narrow but exact tests, using gauges, dials, lenses—extensions of the eye. The small parts stand for larger wholes, whose further juxtaposition then supports the theory or overview of an entire realm.

It is thanks to abstracting that numbers and other symbols can be used; in other words, that measurement is possible. For unlike the coarse phenomenon, the abstracted element or unit is always uniform and the counting of similars is legitimate. It was the men of the Middle Ages who brought the conceptualizing urge of mankind to this high degree of subtlety by giving a name to every *conceivable* they could find, and modern abstraction, now often excessive as we shall see, emulates their zeal.*

What modern science added was the new power which comes of expressing the bonds between abstractions by numerical means. Having eliminated the uncertainty of actual things and produced the scientific unit, the inquirer can state in fixed terms what it does. The great virtue of numbers, as Whitehead points out, is that they can be used to carry on long trains of reasoning without bothering about the subject matter.† Numbers are, besides, transparently clear

---

* The doctors were not so foolish, and Molière not so clever, as people think when they laugh at the explanation that opium puts people to sleep by virtue of its dormitive power. To classify opium as a dormitive is to take one step up the ladder of conceptualizing, and toward the inventory of what substances induce sleep.

† One is appalled when one reflects that two thousand years of verbal reasoning would crumble if the word "insofar" were disallowed. "Insofar" is what science supplies a numerical ratio for.

to the eye, which reduces the chance of error; and they are wonderfully compact. When Condillac in his *Langue des Calculs* gave an example of algebra retranslated into words, he used two pages of difficult prose to convey what was originally in two simple lines.

Since abstractions stand for phenomena, and numbers for their interconnections, the natural drift of the mind reared on science (or close to it) is to take the peculiarly abstract abstractions of science for the things themselves and soon come to believe that only what is measurable is real. Thus a famous psychologist, bent on bringing the mind within his scope, declared that whatever exists exists in quantity, and whatever exists in quantity can be measured. Again, scientific as well as common speech refers to genes as things which "do" this or that. Texts on modern physics refer to the four-dimensional continuum as if it were an object of perception which "acts" thus and so. Yet the same scientific logicians who speak in this way also doubt the existence of ordinary things. And since the reasoning throughout rests on analogy (the abstraction is *like* the real thing; the statistical unit is *like* the individual case; the inductive law is *like* a descriptive enumeration), the tendency is to forget the distortion that has been practiced upon experience for the sake of science. The principal distortion is that of selection, for selecting means attending to one thing only, and regarding the rest as unimportant, unreal, nonexistent—at least temporarily. But the temporary becomes permanent by long inattention, and thus it happens that scientific specialization, by growing narrower, reintroduces a kind of subjective bias into its work, each worker seeing nature through "his" phenomena.

This danger, which is steadily resisted as a type of anthropomorphism, or projection of man's shape upon the screen of events, is compounded with another and graver one. Because scientific method excludes from the start whatever cannot be brought to the test of visibility, and constructs a material machine out of selected phenomena, the scientist—once again understandably forgetful—comes to believe and to say that matter is the only entity. His success blinds him and others to the fact that he began by assuming the conclusion which he now presents as having been found and demonstrated. In a word he turns metaphysician, which he has a right to do like any other man, except that in so doing he violates his original contract with reality.* He also puts science in the peculiar position in which we find it today, of half-claiming and half-exercising a monopoly of truth.

The history of this assumption of power is recorded in the vocabulary of abuse which goes with scientific materialism—in particular the derogatory application of such words as: "metaphysical," "mystical," "subjective," and "anthropomorphic." The implication of the last-named is that in a well-poised and well-trained mind nothing should have the shape of man, nothing should correspond to human tendencies and desires; rather, everything should mold itself on the shape of matter and its behavior as taught by science. This amounts to saying that much of experience is to be denatured, or denied its part in the common life.

I am not here arguing the merits of either view. I am describing, in order to understand the present activity of science, the outcome of its long struggle with religious beliefs, moralities, and metaphysics that were unquestionably limiting and oppressive in their way. We owe to the conscious and half-conscious materialism of science our liberation from one sort of tyrant. We owe to it the stance of action, instead of resignation, in the face of material evil; and we owe to the vehement propagandists of that brand of science its establishment as public benefactor, an acknowledged merit it thrives on, regardless of its frequent disclaimers of utility.

The great theorist of such a science was of course Bacon. He was indeed the prophet of our scientific culture, for he laid equal stress on the utility of science and on its power to satisfy man's fundamental need of knowing. Bacon was the first to depict the man of science and the work of science as separate and special, and to demand that the new discipline be organized.

Despite his latter-day detractors, Bacon's great reputation is deserved. He influenced the English experimenters, including Newton, and he taught the French Encyclopedists, who taught the world. It is therefore ungrateful of modern writers to dismiss the first contributor to what is now called the sociology of knowledge,

* As Whitehead points out, "there is a trimness about it [the materialistic theory] with its instantaneous present, its vanished past, its non-existent future, and its inert matter. This trimness is very medieval and ill accords with brute facts." *Concept of Nature*, 1920, 41.

and to scoff at the insufficiency of the so-called Baconian principles. In describing the "democracy" of scientific work, the role of specialization and coordination, and the dangers of premature inferences from what "science says," Bacon showed a creator's foresight. His unimportance as an experimenter—a taunt echoed from his contemporary enemy, Gilbert—is quite irrelevant to his ability as a strategist of science.*

After showing that science must rely not merely on observation but on experiment and hypothesis, Bacon comes to the main tenet of science, the root idea which colors all modern thought and feeling, scientific and unscientific, the idea of the *irrelevance* of man. Purpose, says Bacon, is something which has to do with the nature of man, not with the nature of the universe. Objectivity is obtained in science by recognizing that phenomena are without purpose. Truths are known by their divorce from manlike characteristics; they are positive, to use Bacon's own word,† and the history of later science has borne him out. The use of terms that suggested human analogies has often produced error: "force" implies something muscular and pushing; "energy" is more impersonal. Early chemistry similarly spoke of "elective affinities" among elements; the notion of "combining weights" freed the mind from the manlike idea of predestined loves and led to the discovery of purely numerical relationships and a natural grouping of elements.

There is grandeur in this ideal, this *mystique* of scientific method, which men have been willing to live and die for. These heroes and martyrs of science are not inventions of publicity agents, and their selfless fortitude is all the more striking for being usually silent and long drawn-out, unsustained by dramatic acts or a

---

* For a just estimate of Bacon, see Blake, Ducasse, and Madden, *Theories of Scientific Method: The Renaissance Through the Nineteenth Century* (Seattle, 1960). The authors' conclusion is that "Bacon's greatness may perhaps most truly be said to lie in having written lines between which so much that was true and vital was later so plainly to be read" (70). Indeed, the thirty pages of *The New Atlantis*—to say nothing of the *Novum Organum*—will show the strength of his vision.

† Newton's *"hypotheses non fingo"*—"I don't make hypotheses," which is not literally true, had doubtless the same function of asserting the primacy of the *data*, the *given*, in the reports of natural science.

watching public. In solitude, mind and hand are at work narrowing, enlarging, destroying the conditions under which the stuff of the cosmos moves, and espying the signs of its recoils. Then, after sight of the clue, comes insight into the reason of its being—"the uprush of discovery; the solemn joy as the generalization rises like a sun upon the facts, floods them with a common meaning."[1] Or if the words of H. G. Wells's fictional hero seem exaggerated, we may prefer those of Einstein: "In the light of knowledge attained, the happy achievement seems almost a matter of course. . . . But the years of anxious searching in the dark, with their intense longing, their alternations of confidence and exhaustion, and the final emergence into the light;—only those who have experienced it can understand that."[2]

This tight net, then, of attitudes, methods, concepts, and buried assumptions supports the classical science which continues to inform our speech and thought and much of our laboratory work. The impatience with whatever is not positive in Bacon's sense is a cultural mark of modern man in the West.

But there is also another science, which follows without break or seam from the first, yet questions many of its beliefs and all of its certitude. It is a self-conscious, sophisticated science—an avant-garde. As Whitehead once said that the last thing a science discovers is what it is about, so it may be said of science at large that it is only now beginning to discover its hidden nature. Twentieth-century physicists and mathematicians have been especially concerned to find out the purport of science: what does it actually refer to in its propositions and by what mental acts does it reach them? Historians of science, logicians, and psychologists have joined in the study, which branches out into several views of the nature of scientific thought and its relation to the universe. And while physics and mathematics hold the stage and supply most of the examples, biologists of a philosophic turn maintain that their work, which has an equal right to represent science, does not fit the others' description.

I cannot summarize here the broad debate; the interested reader should be content with nothing less than the original voices. But

I must illustrate the fact of diversity, which is not merely of opinions: science itself is no longer unified, majestic, or what I have called puritanical. Man's shadow falls across the work. It has always been known that scientists differ in temperament, but seldom admitted that they do not follow *the* scientific method, whether by the term is meant a set of rules or a mental discipline. Long ago the French mathematician Poincaré could distinguish types among his colleagues—the logical and the intuitive. Other scientists have ascribed success to intuition, "feel," and hunch, rather than to method. It was left for a contemporary physicist and philosopher, A. A. Moles, to detail with examples from a wide range of past and recent scientific achievements the bewildering variety of procedures that lead to great results.* He distinguishes according to both the temperaments and the intellectual types.

Among temperaments, for instance, he discerns the deliberate characters who might be thought lazy, preferring as they do any effort of theoretical construction to the task of setting up an experiment, from the more active, who are primarily experimenters. Others are tenacious and succeed by dogged lab work in clearing up an entire field cluttered with rival theories and inconclusive experiments. No less useful are the dilettantes, who, being drawn to many topics without pursuing any very far, are of great importance at the beginning of new sciences; Leonardo is a prime example.

The actual methods used in each case by each type cannot of course be fully inventoried; human thought is too fluid to be divided and tagged. But Dr. Moles can point to a score of tactical attitudes that differ, though they may combine in any one piece of work. The "method of contradiction," for example, appeals to investigators blessed with more than ordinary skepticism, perhaps with arrogance. It consists in assuming the opposite of the established doctrine and seeing where this will lead. To uphold the negative requires reasons, which require new facts, which require

* *La Création scientifique,* Geneva, 1957. The author points out the difficulty of the inquiry and that of securing scientists' assent to the result, because they are "professionally trained to conceal from themselves their deepest thought . . ." and they "exaggerate unconsciously the rational aspect" of work done at some time in the past (99-100).

the invention of new experiments. It is a risky method, for doctrines do not get established without the support of innumerable observations. But when it succeeds it means a "revolution."

Quite different is the method of "haphazard experiment," which sounds like a paradox, but has been repeatedly described by such investigators as Claude Bernard and Becquerel. It was Alexander Fleming who said about his discovery of penicillin that he was "just playing about." Experiments begun "just to see" may be the repetition of old ones with new equipment, or the mixing of two well-known series of phenomena just to see if familiar consequences will be modified. The moves are no more directed than a child's fiddling with a toy, but the supervising eye and mind must be of uncommon alertness to notice the significant discrepancy.

As a method, haphazardness is the opposite of the "method of contradiction" and also of the "method of recodification," which looks for new *expressions* of familiar events, in hopes that new *events* will turn up. Still another, the reductive method, consists in separating an occurrence from all its concomitants, in order to make it artificially a new scientific object. Again—to conclude this random selection from a rich array of mental maneuvers—the "dogmatic method" is frequently used to initiate or redirect experiment. It starts with a wholly theoretical construction—a pure invention—made up regardless of any facts to which it may apply, but logically coherent. The scientist then assumes that any logical construct is true and that his task is to find its proper application. Sometimes this method is used when a hypothesis has been framed to cover certain observations which new facts or calculations disprove. The investigator hangs on dogmatically to the logical model in defiance of the known facts, until he finds another set that fits and removes the original puzzle. In this way Born established his reinterpretation of the original quantum theory.[3]

This enlarged view of method changes the meaning of many words about science that people thought fixed: fact, law, rigor, reality, cause, and scientific truth itself. Thus "fact" means indifferently: what forces itself on the attention, what is known intuitively through the senses, what has been tagged to mark an aspect of a system, what has been inferred reliably from observation, what is defined, measured, and recorded.

Likewise with "law," which as early as 1873 Clerk-Maxwell redefined, apropos of molecular action, as statistical knowledge which could not be refined upon to give exact knowledge.[4] The bad metaphor "obedience to laws" was never intelligible in a mechanical system, but its absurdity grows, if that be possible, when it is seen that the individual particle is not bound to act like the statistical majority. The laboratory determinations themselves, which quite rightly impress the layman, do not give unvarying results, nor is exactitude guaranteed by taking the mean of three scale readings: it is a convenience, which is sometimes abandoned in favor of an intuitive choice of the "best" reading.

If these compromises lead to success, as we know they do, then reality can no longer be said to be embodied mathematics. Many workers still interpret slavishly Lord Kelvin's dictum about knowledge being "meager and unsatisfactory" unless it is expressed in numbers.* Relying more on industry than on imagination, these laboratory men will measure whatever they can and fill the literature with their figures. Forgetting that the footrule is but an artifice, and lacking the finder's intuition for the "go" of things, they skip the difficult step of choosing what to measure. But this hangover from an earlier conception of science reminds us that the rhetoric of an age influences science like the other professions. In the seventeenth century the prevailing form of scientific rhetoric was geometrical demonstration, and Newton was constrained to adopt it so as to persuade his contemporaries, though it was not well suited to his experiments. In the nineteenth century what carried conviction was slow gradual change by small increments. Today, the logical model in symbolic form and (as we shall see) the idea of metaphor best inspire the sentiment of rationality.†

Interpretation by metaphor strikes at the root of the former

---

* This is again the physicist speaking for all of science and leaving outside "satisfactory knowledge" the description and classification of the earth and everything upon it.

† It is this prevailing power of history that makes it foolish to regret, as some have done, the Greeks' prejudice against foreigners which held off the use of the convenient Arabic numbers. Greek genius, it is said, could have accomplished as much in arithmetic as it did in geometry. This is to forget that the impetus toward geometry was the desire to overcome the difficulty of an arithmetic using clumsy numbers.

belief or postulate of identity among material particles. A is *not* B, it is merely to be thought of as B. Metaphor, unlike abstraction, relies as much on difference as on similarity. The two halves of the comparison, the like and the unlike, act as pointers: there is no metaphor, its tone and meaning are lost, if one forgets that it points to similarity-in-difference. The recognition of metaphor in science therefore precludes the idea of the universe being a machine of homogeneous parts. This has important consequences for those whose specialty is to discern the nature and function of science. Since the direct gripping and measuring of matter is no longer pre-eminent and laws are but statistical remarks, science no longer says that it knows such-and-such a thing to be thus and so, but that the thing "may be known as—." Science, then, gives a metaphor instead of an affirmation; it reports on what it conceives in the form of a perpetual "as it were." The very word "concept," which has in our time gained vulgar currency, implies "*take* with" in the mind, and not: "receive ready-made." In seizing upon what it studies, science no longer makes assumptions and expects that they will yield verified causes. Rather, it makes presumptions and is not concerned with the cause or the reality of its objects.

For some theorists of science, the upshot is that "every scientific statement remains *tentative forever.* . . . *We do not know: we can only guess.* And our guesses are guided by the unscientific, the metaphysical (though biologically explicable) faith in laws, in regularities which we can uncover. . . . Like Bacon, we might describe our own contemporary science—'the method of reasoning which men now ordinarily apply to nature.' . . . Science is not a system of certain or well-established statements; nor is it a system which steadily advances towards a state of finality. Our science is not knowledge. . . ."*

Einstein had himself found it "difficult to attach a precise meaning to the term 'scientific truth,' so different is the meaning of the word 'truth' according to whether we are dealing with a fact of experience, a mathematical proposition or a scientific theory."[5] But that truths in various forms and for various uses emerged from

---

* Karl Popper, *The Logic of Scientific Discovery* (1934), New York, 1961, 278. See also: Henry Margenau, *Open Vistas: Philosophical Perspectives of Modern Science*, New Haven, 1961; and Stephen Toulmin, *Foresight and Understanding: An enquiry into the aims of the sciences*, London, 1961.

science he does not seem to have doubted. His contemporary Popper is sure of his demonstration that science "can never claim to have attained truth or even a substitute for it, such as probability." Science consists of "anticipations rash and premature and of 'prejudices,' but these . . . are carefully and soberly controlled by systematic tests. . . . None of our anticipations are dogmatically upheld. . . . On the contrary, we try to overthrow them . . . in order to put forward in their stead new unjustified and unjustifiable anticipations."[6]

Whether or not all competent judges reach this conclusion, which leaves as the sole justification of science the quest for an unattainable—and indeed demonstrably nonexistent—Grail, the modernists in the art of discussing science agree that the connection between the objects of science and the stuff of either "reality" or common sense is so remote as to leave untouched the main contention: science deals not with existents but with features of a conceptual scheme.* Therefore the quest can never end. Its aim is to refute, not verify. For all abstractions (now raised to metaphors) being partial, they give rise to contradictions and thus force an endless revision. Reality the Unknown appears to respond to this human effort by "meeting our theories with a decisive No—or with an inaudible Yes"; but that is all. Reality is inexhaustible and, one trusts, the human mind as well.

The sympathetic observer cannot help concluding from what he reads and hears that even though there is but one scientific culture, not two, there are at the heart of it two sciences. For he can see that the avant-garde views arising from what happens in the extraordinary engines of physics do not influence the minds of the many others who slice rocks, study fish and fossils, investigate the starch molecule, pursue research on horses to cure human asthma,

---

* "An electron for us," says Whitehead, "is merely the pattern of its aspects in its environment so far as those aspects are relevant to the electromagnetic field." *Science and the Modern World*, New York, 1926, 191. A comprehensive and also a conservative reflection of the same attitude is found in Ernest Nagel's *The Structure of Science*, New York, 1961, in which apropos of social science the author concedes that "the discovery of the necessary and the sufficient conditions for the occurrence of phenomena, which is sometimes said to be the goal of science, is an ideal rarely achieved even in the most highly developed branches of natural science" (582).

or—like the late Dr. Kinsey—collect verbal reports from the citizenry to find the range of its sexual attainments. Scientific opinion among those engaged in educational and industrial work remains traditional and cares little or nothing about the ultimate logic or illogic of science. These practitioners are for the most part unconscious materialists and impatient of any discussion that would probe below the surface of their working beliefs; that is, the pure nineteenth-century mechanism, modified as to particulars by the awareness that radioactivity and quantum theory damaged the perfection of the great world-machine.

Habits of speech also prolong the coexistence of the two sciences. For example, a philosophical biologist tells us that energy must not be talked of as stuff, for it is but "a function of a physical system, the quantitative aspect of a structure of happenings."[7] This is very well on paper, but when the structure or the happenings or the energic charge or mass becomes the subject of a sentence, speaker and hearer are both tempted to think they are discussing some *thing*.

Again, in the purely abstract realm of mathematics, the kinship of those abstractions with the world and with the mind is kept uncertain by the inevitable rhetoric. While a professor of engineering mathematics says that he is drawn to his work by the beauty of "statements about universes that the mind makes for its own delight, universes that have nothing to do with the physical world in which we live," a pair of mathematical statisticians brush beauty aside to remind us that "the notion that the intellect can create meaningful postulational systems at its whim is a deceptive half truth. 'We are servants, not masters of truth.' "[8]

Meanwhile the master builders of the new social sciences are certainly convinced that they are dealing with reality, however much they abstract from it and try to translate it into mathematics. Almost always, too, their aim is to predict for useful ends, for social betterment, and their zeal would be damped if they felt constrained by the new logic of science to adjust themselves to a quest without use and without end.

The biological sciences necessarily occupy a middle position between these extremes. Their knowledge has many derivative uses, and the strict mechanical interpretation has never worked well in their realm; for it is the realm of things that live and hence seem to pursue a purpose, at least that of surviving and reproducing themselves. At the same time as this fact lies heavy on biology, the chemical study of the chromosome promises to lay open the material workings of the phenomenon called life, the matter in it having by definition no purpose: it is bewildering, especially now that a second scheme of heredity, additional to the chromosome, has been found.*

In any case the biologist wants to retain his status of scientist by being as wary of anthropomorphy, as unwilling to make room for purpose, as objective (he would say) as ever. He means by objective the traditional assumption that phenomena occur without reference to the observer, the assumption which physics has now abandoned, recognizing through its own work an "historical" reality, that in which the knower and the object of knowledge are united. This is not an arbitrary addition but one made necessary by the behavior of particles. Yet it scarcely comforts the observer of Nature (in the sense of living nature), for the apparent concession to man as observer is nullified by the abstractness of the entities discussed. The mind moves away from the world—as it does also in modern art. There is indeed much nowadays for the eye to see, but in both art and science the recognizable is taboo.

It is accordingly difficult to give a single name to the emergent quality of scientific culture at its purest. Taking the historical view that man begins the quest as a feeling creature, groping on earth and with the sky above, what has he done with his surroundings and his opportunities so far? Techne is the answer of his hands. Science is one answer of his mind. The denial of his own consciousness and rejection of his own purpose being its chief invention, what are we to say of its spiritual tone? For man to think as he once did: "I am the consciousness of the universe, hence the center and measure of all things earthly, and the Creator himself watches over me" is perhaps less arrogant than to say: "I am nothing in what goes on. Without wish, hope, or purpose, I stand

* The work of Dr. Ruth Sager and of Dr. Robert S. Edgar, announced in September 1963.

outside all that is, blind in the blind cosmos; but wait, and I shall deliver and interpret it for you piecemeal and then complete."

At first it seems as if the utterer of this prophecy were acting out the very dream of childlike curiosity, which is to look but stay free from control and loose from utility, seeing but unseen.* Yet troubling thoughts break in. The abdication of purpose feels like a denial of the life within. Man is all of a piece and, as one might say, self-gregarious: he finds it painful to get on without himself, to be perpetually self-conscious, and this about an endlessly increasing number of things. He reflects, too, that although objectivity has for him the grandeur of the unnatural and ascetic, curiosity without wonder is a swinish emotion. As he finds out in the realm of techne, so in science (and in art) he cannot indefinitely admire cleverness, subtlety, and technique. He wants substance, and it is no play on words to say that the voiding of substance out of physics and of purpose out of biological Nature leaves him wretched.†

This may be why today such strong passion is shown about life in other worlds, about the origin of life, about making life in the laboratory; and again, for sending men and women circling around the earth and perhaps to the moon, thus peopling the eternal spaces that frightened Pascal and making our weak homely human voice heard in those intervals, which in our concepts we take to be even bleaker than they are. And finally we are perhaps stirred to life by the wholesale power we have of killing life—plants and insects, wildlife, our food, and our fellowmen.

The truth is that we have lost Nature. In handing it over to the men of science, we hoped to receive it back augmented, transfigured, and full of light. They were indeed its faithful guardians for several generations and they still talk of it in their public

---

* It is not invidious but factual to remark that modern experiments in social psychology reproduce this situation by looking through one-way screens at unaware subjects, arranging tests that conceal other inquiries within them, and in general breaking through the ancient barriers of privacy and ethical conduct for the new purposes.

† It has no bearing on these emotions that matter and substance are likewise abstractions and were shown to be so as far back as Bishop Berkeley. The parallel between the abstraction and the stuff of our lives was none the less clear and satisfying; Huxley could lecture "On a Piece of Chalk."

utterances as a unified power which "acts" or "yields secrets" or "responds" to them in the course of their work. But when the name and the capital letter is not an evasion of logic or metaphysics, it is nothing. Nature is but a vast blank screen upon which the experimenter throws his abstractions and hypotheses, seeing no more—as modern man and specialist—than the minute system of his metaphors and his inferences.

It is ultimately from this atomization of Nature that for modern men the vagrant despair comes, and not from the hurts to their vanity supposedly caused by Newton, Darwin, and Freud. And linked with the disproof of this cliché is the undoing of another. I refer to the allegation of religious moralists that man's present trouble is self-worship. If it were so, he would be vainglorious and jubilant instead of miserable, and no demotion by the unholy trinity could dent his self-satisfaction.

What can I mean by saying that we have lost Nature? To all appearances "nature-loving" has never been so widespread, stimulated as it is by the need to flee periodically from the noise and fumes of Techne. Books and magazines serve this interest, as do societies of bird-watchers and other groups of specimen-hunters and view-finders. *Walden* continues to be a best-seller and conurbanites demand a vacation to hunt and fish and sport in nature. Nor are these people in rebellion against science; on the contrary, many think of themselves as modestly participating in scientific work, and they contribute important facts.* But it is significant that these pastimes almost always turn to study and system, and are justified and respected on that account. The detailed learning takes the place of any cosmic view, and both the participants and the books they read ridicule the "romantic fallacy" of finding in nature anything other than the criteria and the pig Latin of classification.†

Compelled by the same Puritanism of science, even our poets

---

* With such competitive zeal that recently the great British institution of bird-watching was shaken to find that many reports of rare species in the home counties are probably ornithological frauds. *New York Times*, Aug. 10, 1962.

† For an excellent and typical attack on the "romantic fallacy," see the chapter bearing that title in Robert Ardrey's *African Genesis: A Personal Investigation into the Animal Origins and Nature of Man*, New York, 1961.

and critics have abjured "communion with nature." They deride Wordsworth, Thoreau, Novalis, and the rest of the Romantic company for raising a hope and not being able "to answer the terrible question of whether, once they committed themselves to the proposition that their most delicate experience was typified in Nature, they were thereafter actually writing about Nature—about Walden Pond, for instance—or about nothing more than their delicious experiences."[9]

The question is terrible in part because the modern skeptic knows only at second hand what Wordsworth and Thoreau said. It is taken for granted that they were feeble-minded in their belief that Nature was altogether benign and gentle, for it is one of the by-products of the march of science to assume that earlier men could not see what lay before their eyes. The truth is that Romantic writers beginning with Rousseau did not idealize nature; Thoreau in particular knew that it was more to be admired than deemed hospitable to man. But they knew from experience what Wordsworth made poetry and doctrine of, namely that the habitual association of one's ideas and feelings with a particular spot of earth or aspect of nature enhances the dignity of human nature by engendering tranquillity. And we must remember, when we think the only tranquilizers come from the chemist's, that this doctrine going back as far as Rousseau was already a rejoinder to a mechanistic materialism and a zeal for analysis and abstraction—the work of the Enlightenment. In logic, the Romantic doctrine is as tenable as that of the latest physics, which projects an idea upon aspects of a scheme and tests it for congruence. Wordsworth chooses his aspect and records his experience, which is a test.

But before pursuing this analogy, one must recall the grim destiny of nineteenth-century science, for it is not merely the Wordsworthian emotions that were swept away, but their substitute as well. Science offered mankind a vision of order. Ubiquitous science professed to outshine the poets and the theologians and the philosophers at their own game, and by this promise it succeeded in overthrowing them and gaining power over all minds. The new scientific order was called Evolution. It was a cosmic philosophy resting solidly on previous work in physics and chemistry, and at last able to confront its public and its enemies with the face of living Nature. Evolution, the century was told, comprehends life and matter in one scheme; it works mechanically, without purpose, just like the stars—and yet behold the manifold beauties this mindlessness has produced! How much grander and more credible than the stories handed down from the devotees of a tribal god. An infinity of worlds, of forms, of processes—each elaborate and fine in its perfect adaptation, a wondrous spectacle, and yet reducible to the simple actions and reactions of senseless stuff.

Note that the propaganda adopted the previous poetical mood of admiration and love of nature.* Humboldt's great synthesis, *The Cosmos,* was largely published when Darwin's book capped half a century's evolutionary doctrine, and *The Cosmos* was then science's response to the desire for an accurate and imaginative panorama of the physical world. Evolution coasted, as it were, with the momentum of this desire and of the educated public's knowledge of natural history. The new golden gift was a unifying principle that was the same for man, the earth, the Milky Way, and the animal kingdom. Immediately and feverishly, new accounts based on evolution were given of every institution, law, custom, art, social tendency, and human trait. For forty years the cosmological passion was insatiable.

But evolution failed. It betrayed its believers on its own ground and outside. First, the social and philosophic applications collapsed because the explanation by accident and survival would not fit the history of any human enterprise. In biology, evolution does indeed subsist, but so modified by the new science of genetics that it can no longer claim the merit of a clear, simple, total explanation which it possessed for the late Victorians.

Moreover, it lost their faith and their descendants' on the point of credibility, which is negligible in science, but all-important to the cosmological appetite. Granted that separate creations and a foreshortened geologic span are not believable, neither is it believ-

---

* As Robert E. Fitch has shown in the representative instance of Darwin himself, the propounder of Natural Selection was an unashamed nature-lover and did not resemble the severe image of "the scientist" now drawn of him. "Darwin and the Saintly Sentiments," *Columbia University Forum,* Spring 1959, 7 ff.

able that pure chance could combine atoms and cells into forms
that fulfill a profusion of complex and varied purposes. The words
"natural selection," "adaptive value," "variation," "environment,"
"struggle for life," and "survival," now seem question-begging
formulas that do not mask or answer the grave difficulties which
from the start made Darwinism unsatisfactory.*

As a bootstrap idea in the biological sciences the usefulness of
Darwinism is so manifest that it would be as foolish to impugn it
as to impugn the materialistic creed of the worker in the laboratory.
The one is the extension of the other and both serve by making
the sciences of life hang together, without getting in the way of
particular determinations.

It is another matter when the lay believer is asked to take
Evolution outdoors with him and live with its implications. For
one thing, the explanation of origins which he ponders is not fur-
thered. For another, the spectacle of nature continues to impress
his mind as much with fitness and purpose as with blindness and
chance; and a good part of the fitness appears as the fitness of the
environment.† Accordingly, a modern imagination predisposed to
a belief in science, yet trained by art and experience to be subtle
and aware of itself, will generally find that neither creation nor
evolution overcomes its profound conviction of ignorance.‡

At this point the troubled cosmos-seeker may be told not to
imagine at all but simply to conceive, to think "as-if." If he con-
sents he must at once ask "As if what?" For the answers to his
request for a livable Nature vary widely. Some will tell him that

* Darwin himself was not satisfied with it, and in his latest revision of the
Origin he propped Natural Selection with two other causes long in the field—
one first advanced by Buffon, the other by Lamarck. See the sixth and last edition
of *The Origin of Species.*

† If he is a close observer he cannot help thinking that structures such as the
eye, functions such as the reproductive apparatus of mammals, require so many
cooperating parts and secretions and constant internal adjustment as to defy the
likelihood that a portion of the evolving form could wait for a lucky variation
to occur and be saved to lend its assistance. And in the development of man, what
survival value has music, hairless skin, the lower testicle? The introduction of
genes to do some of the work only hides the difficulty.

‡ See, for example, what Joseph Wood Krutch, the well-known desert naturalist, says of the difficulties he feels about the standard explanations of the origins
of life, sense perception, and "contingencies of the universe." *The American Scholar,*
Winter 1957, 100-102.

biology cannot give meaning to the statement that an organism is alive. The very business of that science is to show that physico-chemical materials give an appearance of life to what is "really" as dead as a doornail.[10] This is "as if" with a vengeance. Another will say that it is no longer possible to feel as if congeniality existed between man and nature, because in his latest efforts to understand nature man has to build himself a barrier against lethal radiation.[11] This is to play on the word "nature," which has become more slippery than ever since "the natural" is being added to by human means.* Still another asserts that "in passing from physics to biology one is conscious of a transition from the cosmic to the parochial, because from the cosmic point of view life is a very unimportant affair."[12] This is the language of the senior recruit, toughened and rejoicing in the civilian's pain. And it is a superiority without basis, for it presupposes a "cosmic point of view," which by definition does not exist; there is only the point of view of man as scientist expatriating himself from the cosmos.†

The crux of the difficulty for both scientists and laymen would seem to lie in the ambiguity of the word "order," which they seek alike for the satisfaction of their respective emotions—the emotion of the worker and the emotion of the contemplator. For science, what happens in a plane crash to bodies living and inanimate is an example of perfect order. No severed head rolls one millimeter farther than its mass and the vectors of its course define. To see in this a universal order and to love its immutability is an idea worthy of man's imagination. It was first deeply felt by the Greek dramatists under the name of Necessity. But order has a second meaning, the meaning of purpose, which the plane crash violates. The accident, as we rightly call it, produces disorder in human affairs and affections, and that conclusion is as absolute as the other.

* A reporter writing on a recent satellite finds himself led to say that the device "for the first time uses only the forces of nature to keep one face always pointed at the earth." And the headline speaks of "New Principle" (*New York Times*, July 23, 1963). One sees what is meant, but the implications are none the less hard to sort out.

† As Whitehead remarks, "Scientists animated by the purpose of proving that they are purposeless constitute an interesting subject for study." *The Function of Reason*, Princeton, 1929, 12.

This is why, since 1900, biologists whose specialty or whose temperament brings purpose vividly home to them in their work try to bridge the gap between the blind scientific order and the purposive order by introducing into their scheme the element of pattern or form.* They combat those of their colleagues who assert that the cell is the unit comparable to the atom, from which it follows that the organism is not a unity. They retort that a kidney cell extracted from the body is wrongly called a kidney cell, "for it is no longer its characteristic functional self, which we are trying to study."[13] This is not to deny the material home of life but to remember that it has a shape, an indispensable order. This is even true of the inanimate, as the biologist goes on to illustrate: "Iron made into a poker is still analyzable chemically, but it is no longer just a lump of iron explicable solely in chemical terms. It has an organization beyond that of the chemical element."[14] On this perception our century—and particularly the philosophizing biologists—have built their faith in pattern, their *mystique* of form.†

The trouble—if it is a trouble—is that form and pattern are not objective and material. A single pattern can be seen correctly in two or more different ways; and even if not so seen, it remains something of a different kind from what it is a pattern in or of. The struggle here, between the uncompromising materialist and the no less hard-headed observer of the fact of form, has an epic quality for all its appearance of a quibble. The materialist himself has lately shown signs that he is not insensitive to the other's argument. The new way of talking about "the genetic code" and the cells' "carrying instructions" to other parts of the genetic apparatus

* See, for example, the writings of John Scott Haldane, L. J. Henderson's *The Fitness of the Environment* (New York, 1913), and Sir Charles Sherrington's famous work on *The Integrative Action of the Nervous System* (New Haven, 1906).

† Mr. Lancelot Whyte is among the best known. His book *Accent on Form* is virtually summarized in one sentence from it: "We can look forward to a unifying philosophy of form, displaying wherein we are one with all nature and wherein we are uniquely human." New York, 1954, 101. More cautiously, J. H. Woodger says: "It will be necessary to devise ways and means of obtaining data requisite for correcting a purely analytical procedure." *Biological Principles*, London, 2nd impr., 1948, 315.

is obviously a means of avoiding the old mechanical analogy, though the language also owes something to the vogue of computers, which are machines. This new mood is to be noted as an effort to break away from the deep-rooted habits of our minds, derived perhaps from the original Stone-Age billiard ball. We find it almost impossible to conceive action without sequence. We always imagine one thing coming first and pushing the rest—Heredity *or* Environment, the gene behind the trait. Interplay—simultaneous, reversible, inconstant—we cannot grasp with common sense; and it cannot be condoned by science, despite all the electrons in captivity, if science is to keep its order of accident, rejecting the order of purpose.

Does this continuing debate between the Two Sciences (or more) not warrant the "aspect-seeking" desire of the nature-lover whom I left in suspense some pages back, in order to retrace briefly the saga of Evolution? Perhaps it does, but there is another obstacle in the way of the modern imagination: it is that other chapter of the evolutionary story, heartless nature and animal man. The public has learned too well the cliché of the animals' endless, desperate, fearful struggles, which man must emulate to survive since he is one of them. The sum total is a cruel world of nature, red in tooth and claw.

This notion fitted the time and place of its origin: industrial England, but this anthropomorphic taint is not yet acknowledged, let alone washed away. Nor has it been noticed that the idea of cruel nature must be a fallacy in a system that excludes purpose.* It is to force human judgments on neutral facts when the snake is called cruel because he eats a frog. Would not his starving be cruel too? We do not know how the animals look upon this, or for how long at a time. If we are animals and love danger, as

---

* One of the most instructive blunders on this point is the passage where John Stuart Mill, speaking for the scientific outlook in his essay on Nature, is so outraged by what he believes to be Rousseau's personification of Nature as benevolent that he points indignantly to Nature's well-known habit of criminally killing each thing that lives; he is so indignant that he goes on to point out that *"Nature does this once"* to every being (italics added). *On Liberty and Other Essays*, New York, 1926, 159.

we evidently do, maybe the rest do too: why should they be pitied at the same time as we maintain that they have no purpose and are virtually dead matter whirling in the void?

Nothing could better show the split in the modern soul between spontaneous feeling and late-acquired learning than this dogma of heartless nature. Nature is *indifferent,* no doubt, in the aspect taken by the physicist. But nature along the brook, the shore, or the wood path has features congenial to man. It responds simply in being good to look at, just as it does in giving fruit to eat. When this is so, why should man and nature be radical antitheses? For man, or at any rate, *to* man, nature is colorful, animated, rhythmic to the ear, varied to the eye, solid to the touch, and stable underfoot; it is edible, drinkable, sittable, swimmable, and full of surprising fragrances, tastes, events, processes, and companions. That nature contains other less pleasant things does not change or destroy what nature proffers. Yet we are supposed to say that it does. "The more I see of life," writes Cyril Connolly, "the more I perceive that only through solitary communion with nature can one gain an idea of its richness and meaning. I know that in such contemplation lies my true personality, and yet I live in an age when on all sides I am told exactly the opposite."[15]

What he is told is not so much that he dare not feel and acknowledge what he feels, as that he dare not think except in a certain way. For while man gratifies his senses in contemplating nature he also takes in the size, force, distance, multitude of things; he is penetrated by the needless profusion of life, moved by the forever untraceable destinies of all that breathes, feeds, flows, erodes, blossoms, hibernates, or dies. In those moments man *is* the center of the universe; for being perhaps solitary and certainly silent, he is the conscious focus of Being. And he more gratefully ponders his own.

Though incomplete, such moments bring joy even when they are overwhelming, because they hint of completeness and moderate the self. In this, nothing can take their place, except art drawn from the same source and on a like scale; whereas science, which a hundred years ago gave signs of supplying the grand interpretation of this order of nature, was forced by its logic into opposite ways, with the result we know. Our perverse lust to see, our *voyeurism*, leads us to divide and subdivide with a rodent's tooth and to search out with a Cyclops' eye the ultimate grains of reality, until we grow blind, and having lost nature,* see only the luxuriant abstractions born of our enormous conceptual powers.†

The layman, like a disappointed stockholder, finds it hard to see anything but a deficit. As Auden muses "After Reading a Child's Guide to Modern Physics,"

> This passion of our kind
> For the process of finding out
> Is a fact one can hardly doubt,
> But I would rejoice in it more
> If I knew more clearly what
> We wanted the knowledge for—
> Felt certain still that the mind
> Is free to know or not.[16]

Despite the confidence of some respected elders, science no longer promises a vision of order.‡ Latterly its many motions and rapid decay have been as discouraging to the interest in cosmology as its increasing inaccessibility. An intelligent man of today, even one trained in a science, is as likely to understand the system of the world as a cat the car under which he sits. The thinkers among practicing scientists are themselves unhappy at the complication and excessive ingenuity of the subatomic scheme and its so-called particles.

* In his remarkable book, *Solitary Confinement* (London, 1961), Mr. Christopher Burney points out that a naturalist like Gilbert White could not be made up for by adding together a hundred specialists. "The parts are not quite equal to the whole" (123). And he subtly suggests that "the size of things known seems to decrease as the number of things known grows" (*ibid*).

† Occasionally a thinker suggests, though feebly, diagnosis and cure: "Something," says Jean Wahl, "has destroyed in a certain measure the feeling of our kinship with the universe (which poetry has retained better) . . . [and] . . . we must give a new, less conceptual tone to these ideas of infinity, time, and matter." *The Philosophical Review*, May 1945, 200.

‡ See the words of Erwin Chargaff above, pp. 81-82. In contrast, Heisenberg would have us see in science a sublime symbolism parallel to poetry and religion; and Einstein, at least in mid-career, affirmed that "all of these [scientific] endeavors are based on the belief that existence should have a completely harmonious structure. Today we have less ground than ever before for allowing ourselves to be forced away from this wonderful belief." *Essays in Science*, trans. from *Mein Weltbild*, New York, 1934/1950, 114.

As for the promised control of nature, it is in rout before nature unleashed. On top of this, the deadlock of the two sciences is unbreakable, since the scientist needs for his work an idea at variance with the subtler reading of his results. And for the last, the rejection of man—the rooted race prejudice against *anthropos*—produces in man the malaise to which many testify. It comes not alone from the presence of the Id or the memory of the Ape, but from an obsession with analysis, abstraction, and self-distrust.

If this is the situation on earth, what in heaven or hell is science for? I called it earlier man's most stupendous and unexpected achievement, yet if one accepts the conclusions of the most advanced and hypercritical physicist-philosophers, science has no stability and yields no truth; it has no utility and deals with anything but the concrete. Besides which, it now feeds and inspires techne to exacerbate our desires and torment body and mind. What can justify the word "achievement" for an undertaking seemingly so perverse and prolonged?

The sufficient answer is still Faraday's when he was asked of what use were his laws of electrical induction. He replied, "Of what use is a child?" Out of man's mind in free play comes the creation Science. It renews itself, like the generations, thanks to an activity which is the best game of *homo ludens*: science is in the strictest and best sense a glorious entertainment. And I intend glorious to mean at once magnificent and deserving of glory.

The activities of science are properly a game, because in it convention and chance divide the interest: phenomena are as may be, but the rules restrict their handling, no favors asked or given. Science is play because of the very meaning of free inquiry: not this or that urgent result, but *laisser faire, laisser jouer.* The word "findings" misleads unless one thinks of them as of the temporary close of a match to be later resumed. As Karl Popper made clear in the passage I quoted, it is the quest that matters, seeking not finding. True to the Faustian ideal in its elevated form, science seeks only to know; it is as pure as any human effort can be. It directs itself along any imaginative path, often one of negation, in order to make new—satisfying thus man's deep instinct of anarchy and perpetual revolution.

Through that ever-open door the scientist has escaped at will from the contingencies of the world to those of nature, making that innocent freedom his protecting principle. When Professor Dingo, the geologist in *Bleak House*, was arrested for defacing a public building with a little hammer, his classic defense was that he knew of no building save the Temple of Science. Nowadays, it is true, moral burdens have fallen on some scientific shoulders, but society has made up for this by richly appointing the professor's temple. The magnificent play of scientific mind goes on at a thousand shrines the world over; and by tacit consent and official sanction, the glorious game is declared the most important activity of the nation.

That it is free play is insisted on by the participants whenever they define their work as pure and independent research, that is, untouched by utility and undirected from without.* Beneath the play of mind there is in science the direct pleasure of manipulation, of fiddling and trying out. Curiosity in the young is first a restless physical urge to touch and tumble things together. Energy being limited, it may nowadays exhaust itself in this fashion, leaving little for other modes of curiosity. At any rate, though past ages deemed it low and prying to look into things and be adept at handling them, today it is a pleasure and a source of pride to many. The perfect playground is of course the laboratory, where one makes equipment, adjusts delicate gauges and balances, tips bottles, solders wires, and releases power or reagents in an august form of cookery. How infinitely more engaging than to read books! And outdoors the methods take on the congenial form of the hunt: to establish that porpoises have remarkable "sonar apparatus" men lower fish of known size soundlessly into pools, behind a baffle screen or in the dark of the moon. After which the reports, spare of words, sustain the sense of ritual with symbol, table, and graph. Endless are the forms of calculated play: to send

* After the American astronauts, whose main task is to find out how man can live in space, had brought back all the data for which they circled the globe—data which the public regards as scientific—a set of pictures taken by Major Cooper with a hand camera during his sixteenth orbit suggested a clue to the phenomenon of airglow in the so-called "junk layer" of the atmosphere. "Nothing else done by the astronauts," said Dr. Edward Nye, who is studying these matters, "falls into the scientific category." *New York Times*, Aug. 12, 1963.

up a fifty-pound balloon and surprise the formation of high-energy electrons in space; to interrogate bats by mute indirect ways and seize upon their mode of piloting themselves; to guide invisibly along underground tunnels possibly nonexistent particles which yet have names and leave traces by darkening silver emulsion on a plate; then gaze endlessly at these needle tracks till a combining thought arises in the mind; to juggle lights and prisms and beams and so make a spectrum of light no man can see; to dig little holes in the floor of the ocean and find out what lives there or for how long the ooze has been; to eavesdrop on the musings of grizzly bears by putting microphones around their necks, having first rendered them docile by piercing their hide with a relaxing dart; to make and load and wear garments with which to simulate fatigue under conditions permitting the measurement of its remotest effects; to make a pair of glass buckets whirl at a great rate on a machine and, having filled them with excreta, hope that an unfriendly virus will isolate itself—these are but a few of the doings of contemporary science that cannot help enchanting the mind of ludic man.

Nor should this reminder that the play instinct is at work suggest belittlement. On the contrary, it brings out the natural bond between scientist and ordinary man and shows how primary is the urge to do the work of science. We have Newton's authority for it in the famous words with which he reviewed his amazing career. Like a follower of Karl Popper, he had, he said, been like a child by the shore, "diverting himself in now and then finding a smoother pebble or a prettier shell than ordinary, while the great ocean of truth lay all undiscovered before him."[17] A scientist of our own times, a Nobel Prize winner and a man high in the councils of government, has given a complementary view of the quest. "I like to compare scientific research," says Glenn Seaborg, "to mountain climbing in an unexplored range. Considerable preparation, training, and strong motivation are required to get up to the upper altitudes even if no one particular stretch of the way is particularly difficult. But once there, it is relatively easy for one to see vistas or even to stumble across new riches."[18]

To gather pebbles by the ocean, to climb mountains for the view, are the searcher's similes and pleasures. The inexpert be-

holder takes part as at a play: "For the first time," writes a citizen
to his paper, "many complex projects, plans, and goals that com-
prise Apollo have been welded together into a coherent dramatic
story so that the public can comprehend what the moon program
is all about. And what a wonderfully bold and challenging program it is!"[19]

But there is yet a third aspect of the great game, which the
word "play," taken in two senses, brings to mind: the comedy of
science. Perhaps our scientific culture is still too young to appreciate it, but it is there to be seen by anyone who at times stands
off and looks at his species. Life turns to comedy, suddenly, when
one takes a second glance at those creatures at work behind a
grillwork of tubes and rubber hose, gravely intent upon a speck
of matter or a flicker of motion, wearing the expression of gifted
squirrels watching for a nut. They are the eager beavers of common speech, and their manhood (womanhood too) is not en-
hanced but made comic when we know that the object of their
passion is the quarter-million-dollar mouse, that the day-and-night
anxiety of their lives is whether the subject's thumb will jerk or
not.*

The very distorting of nature into the minute or gigantic per-
formances we make it do—a spark, a bomb, a precipitate—contains
a comic element, as does of course the hubbub of voices competing
in explanation. The cosmic maker's view, as Milton saw, is humor-
ous:

> . . . he his fabric of the Heavens
> Hath left to their disputes, perhaps to move
> His laughter at their quaint opinions wide,
> . . . how they will wield
> The mighty frame, how build, unbuild, contrive
> To save appearances.†[20]

* The counterpart in techne of the passage from the ludic to the ludicrous in
science lies in the occupation of the testers of consumer goods, an earnest band who
take the boasted product of manufacture and advertising and proceed to tug at its
joints, simulate its fatigue, rig up dummies embraced by car safety belts and hurl
them forward, drop encased typewriters on concrete slabs, and generally enact a
mad rationality.

† "Save appearances" is here a technical term of contemporary science, which
refers to the role of hypotheses in fitting observed phenomena.

And Swift after him made fun of the projectors in Laputa for their abstraction and uselessness. Swift was of course a partisan of techne for abundance' sake, yet he ridiculed the endeavors of some on his island for trying always to beat Nature at her own game. Disregarding ridicule, we too love to play the ape to nature, and in our eagerness are seldom very far from the Laputans' attempt to extract sunshine from cucumbers.

The Swiftian example brings us back with a start to the practical problem confronting the thoughtful—how is science to be *taken*? The splendor of the quest will continue to raise in every mind an uplifting emotion of mingled pride and awe,* but how are we to handle our other emotions—of puzzlement at the coexistence of the two sciences, often in the same mind, sometimes in the same paragraph; or of humor, as when a distinguished astrophysicist speaks of the problem of communication between our own civilization and one which may be located on some planet associated with a star many light years away, the joke here being in the word "problem": "The possibility of such communication by microwaves has already been discussed and efforts made to detect, by radio telescope, possible broadcasts which might be in progress to attempt to establish communications with us."[21]

And there is finally the emotion of simple, sensible fear, as when one of our futurist experimenters tells us not to worry about the devirilization of man and the defeminization of woman, because needful revolutions in sex are in the making at his laboratory. We can have generation without paternity, propagation by cuttings, test-tube pregnancies, and for kicks "chemical adultery, a form of command over the sexual instinct. Man will become artificial, scientific, biological man."[22] Anywhere but in science, that final adjective would rank as sardonic wit.

At this point some observers suggest that science is a form of art

* No one understood this better than the men of the Enlightenment or expressed it more movingly than Voltaire in the passage in *Micromégas* (1752) where the puny mortal answers accurately the cosmic questions put to him by the scornful visitor from Sirius. For modernity, compare this with the account Whitehead gives of the Astronomer Royal's report to the tense audience waiting to hear whether the photographs of the eclipse confirmed Einstein's prediction that the rays of the sun would be bent. *Science and the Modern World*, 15.

—or art a form of science: both are outgrowths of the play instinct, both tax the highest powers of the mind, and both act on mankind in ways beneficent and terrifying. Certain scientists, moreover, feel within themselves powers other than rational which after much groping project the new idea as if out of the unconscious. In many circles it is a cliché that "science does not differ essentially from poetry." Is this, then, the answer to our disquiet, that science should be taken like art, as a mysterious, disturbing grace, a higher unpleasantness ultimately good?

No doubt, as I shall argue later, the human mind has only one fount of inspiration and its soundings feel much the same, so that the scientist can well claim irrational powers like the poet. But thereafter it is bad metaphor to say, as did one mathematician who studied the psychology of creation in his field, that science is "choice governed by the sense of scientific beauty."[23] Beauty here stands for the logic and neatness of an organized whole and the pleased surprise that inventiveness always causes. But mathematics and science in general lack the sensual element that is indispensable to art. Something to see, hear, touch, mouth over—the very palpableness that the latest science has given up constitutes the difference between art and it.* I do not doubt that science gives off emanations beyond its bare bones; a man's works can have a family likeness, as in art. But no one who knows anything of art and of science can conceive of an aesthetics being set forth about a branch of science or about the work of a particular scientist. Any criticism of them that had validity could only be scientific criticism. It is accordingly sheer affectation to repeat the nonsense that Bertrand Russell utters with his usual aplomb: "Mathematics possesses not only truth, but supreme beauty—a beauty cold and austere, like that of sculpture, without appeal to any part of our weaker nature, sublimely pure and capable of a stern perfection such as only the greatest art can show."†[24]

* This is not to say that art has not been trying to give it up also. See below, Chapter X.

† A different kind of confusion—"it does not matter a bit if we call flying an art or a science: the art of house-building is practically one with the science of housing"—is frequent among architects and engineers. For a full-blown example, see the essay from which I have just quoted: W. R. Lethaby, *Form in Civilisation*, London, 1922, 37.

The very opposite would be more to the point of modern man's bewilderment. A general recognition that science is at the other extreme from art and cosmology; that at the least a double view of nature is permissible; that much of life is to be dealt with anthropomorphically, because *anthropos* is the customer to be served—this might begin to clear the ground of errors and stumbling-blocks. Certainly it is a paradox that the passion for unifying all ideas and impressions under the rubric of science, one and indivisible, should have become absolute, when two sciences, one strictly deterministic, the other permissive, have between them given us three worlds for one: the world of common sense, the world of scientific objects, and the world of unknown reality. To claim the right to a double view is modest and not arbitrary; it imposes itself daily on our minds, as Huxley once confessed. He had tried hard, at sunset, to see the phenomenon as earth moving and not sun sinking. He could not: the sun *sets*. Science in this case enables us to add a vision, not to replace it.

But the modern sensibility requires more than this. It needs respite from the increasing complications of analysis and abstracting, and for this, strange as it may seem, the mind and the senses must be given back to the natural *complexity* of things. *Complication* differs from complexity in being an elaborate structure put together out of what has been made artificially simple by analysis —as when we say, "Let X equal . . ." This, we know, is probably the beginning of a lucid demonstration, and it is on this account that Polykarp Kusch is accustomed to telling his students, "If it isn't simple, it isn't physics."

The science—any science—is intricate; but once stripped, the parts show their simple natures and set linkages. A clock is simple in this way, though complicated, whereas the character of one's friend or the poetry of Shakespeare is *complex:* neither of these two can be taken to pieces like the first, or thoroughly understood. Science, in short, substitutes the complication of its system for the complexity of the world, and what man needs is renewed contact with the world.

For he is inquisitive in a larger sense than is fulfilled through the curiosity of research. Where and what am I, whither bound and for what ends? These questions that man keeps asking, all agree that science cannot answer. But the confirming cliché—Science tells How not Why—is falsified in reality by the appearance of answer-giving which science has been guilty of for over a century. And when it has not so transgressed it has issued prohibitions against answers given by others. The truth may be that it is in fact impossible for the scientist—or for any man—to describe without interpreting, and interpreting is only a hairsbreadth away from explanation.

Modern physics, for example, is no longer sure of the "heat death" of the universe so popular with nineteenth-century scientists and so lowering to the spirits of their audiences. But this change has not come from a new sense of responsibility about the moral consequences of metaphors and hypotheses, and it is evident that our fresh metaphors in cybernetics and biochemistry are producing a like effect. With each style of science goes a different stage setting, and from each stage a different dialogue with the audience. Some scientists, lately, have come to feel that the audience regards them with suppressed hostility. They naturally resent the attribution of the bomb as if they alone were responsible for it and not the whole world. Again, the noticeable recoil of the public away from the products of techne and toward those of art is disconcerting to scientific men, and so is the windy praise of the humanities that is heard from people who fear science. But there is a nearer cause of the general discomfort, which is the failure of science to make room for anything else, and this at the very instant when it says its pretensions are small.†

Nor does science, true to its revolutionary character, consider present harm of any importance compared to "ultimate" knowledge. I have nowhere seen any acknowledgment that it was an historical calamity to have saddened, precisely with the "heat death" story, the three generations of men spanned by the lives of Tennyson and Henry Adams. Mental suffering is a fact, as important sometimes as the physical which it often provokes. And to have told such men that morals, beauty, love, art, were only illusions unworthy of intelligent men is perhaps less excusable than the "lying legends" of

† See the shrewd article by Robert K. Merton, "The Ambivalence of Scientists," *Bulletin of the Johns Hopkins Hospital*, February 1963, 77 ff.

primitive or medieval times, which the men of science made such sport of.

It goes without saying that I am not suggesting a censorship of science, nor any suppression by scientists of the philosophic ideas which they have perhaps a greater right to entertain than less searching and questioning minds. What is wanted is a "public science," more self-aware, less brash and futurist in its pronouncements. Public criticism should help keep the tone of speculations speculative, instead of letting it be confident, dogmatic, and culpably metaphorical.*

Unfortunately, the conditions of our scientific culture are not favorable. It is hard to imagine an annual Feast of Fools deriding our mentors as was done within the medieval cathedral itself. Democratic life, on the contrary, heightens the need for status and prestige, and the benefit of clergy which the scientist rightly enjoys goes so far as to make him virtually invisible on the scene where politicians, authors, and even Supreme Court justices stand up to be shot at. The prestige and the protection are in truth so great that they attract many to science and many more to its now numerous imitations. From the search for truth, which only lately became a profession, our culture sees multiplying unchecked a horde of wandering friars with projects in their scrip—lay brothers of this or that ology, who with voluble candor seek its entertainment as well as its glory.

* Whitehead makes a broader demand to the effect that "the ultimate arbitrariness of matter of fact from which our formulation starts should disclose the same general principles of reality, which we dimly discern as stretching away into regions beyond our explicit powers of discernment." *Science and the Modern World*, 135.

VI

The Cult of Research and
Creativity

THE appeal of science to the many, young and old, does not of course reside wholly, or even largely, in the prestige it confers. It also confers power, and of a kind that is not suspect because it derives from Truth. However defined or criticized upon its latest frontier, Science *has* built a world of ideas and symbols parallel to the world of brute matter; it does manage to subject its abstractions to the test of the eye so that wrong conceptions get disposed of; it continues to show a tireless zest for self-correction and the ever-enlarging art of mathematical expression. The good that comes of this disinterested intellectual pastime and reaches nearly everyone is this: because science maintains that the source of a truth is nothing and the method of reaching it everything, no conclusion is beyond doubt; questioning is an honorable occupation; and erroneous ideas and false imaginings are continually being challenged.

Practice is not so perfect as theory, but enough results follow to sustain the theory of pure truth-seeking. Findings, which may be good in themselves, serve an even greater general good in helping to keep at bay the fanciful and menacing from public and private life—fears and spells and old wives' tales.* To this extent science has effected a purification of the common mind. Despite astrology and a few negligible sects, our scientific culture is at one in agreeing that for any important assertion evidence must be produced; that

---

* The difficult but successful resistance to the moral and religious censoring of books in the United States is due in part to the scientific and in part to the legal conviction that "the evidence" must be listened to.

prophecies and bugaboos must be subjected to scrutiny; that guess-work must be replaced by exact count; that accuracy is a virtue and inquiry a moral imperative. To this hegemony of science we owe a feeling for which there is no name, but which is akin to the faith of the innocent that the truth will out and vindication will follow. In its purest form science is justice as well as reason.

By thus submitting the work of mind and imagination to a pub-lic jury, science also offers a working role to anyone willing to apply himself. Not great genius but solid ability is needed, and a strong moral attachment to the method. Carelessness in such an enterprise is as bad as dishonesty, and dishonesty is treason. The irresponsible character frequent among artists cannot be tolerated in scientific work, but by way of compensation, as Bacon pointed out long ago, science can make use of the middling talent. Science thus becomes a vast collective undertaking, whether its practitioners work separately or in teams. And as the range of inquiry has wid-ened to other realms than nature, more and more want to share in the pleasure and credit of this unified activity.

Out of these good impulses and sound reasons has come the epidemic cult of research. If someone today tells another in passing that he is "doing research," the one addressed will almost invaria-bly exclaim: "How wonderful!" or at least "How interesting!" Then, possibly, he or she will ask what the research is about. This attitude is so usual as to seem inevitable, but fifty, or even thirty years ago, the idea that doing research is inherently wonderful or interesting was not a commonplace, nor would the question "Re-search in(to) what?" have come as an afterthought or been over-looked altogether. The change is due to the recent canonization of the name of research, which by becoming familiar has bred a univer-sal faith in the supreme value of the results. In any profession noth-ing can be more important than research. To suggest that practice, or teaching, or reflection might be preferred is blasphemy.

History records no precedent for this extraordinary addiction, unless it be that of the Middle Ages to pilgrimages. When it prevailed, I doubt if anyone who was impelled to save his soul by going to a distant shrine was ever restrained by his friends. He would take off his shoes, pick up a stick, and go off with everyone's blessing, just as today the research-bent abandons his occupation, picks up a box of index cards, and is on his way with shining eyes and a two-year grant, amid general admiration. The very phrase "do research" shows that it is the act, not the goal, that matters; and though not many think of research as saving their souls, society at large does believe there is salvation in it.

The United States is, in this regard too, the land of the pilgrims. They go forth within and without. Government, business, industry, private wealth, the foundations, and the universities spend money without stint for whatever bears the name of research. This example is influencing the world. The demand for researchers in all fields outruns the supply, and much of the directing of research consists in recruiting helpers. Almost all professional organizations have a research branch. As newspapers tell us, centers and institutes for research are established daily, the name of the director being usually easier to recognize than the purpose of the research. It is a legacy from science that at first the goal should be as broad as the universe. The later narrowing is heuristic, from *heurisko*, "I find"; which shows how sanguine the pilgrims are on setting out.*

Their faith had its beginnings in early modern times, when exploration gripped the imagination of western man. The first research institute since the ancient one at Alexandria was the establishment of Prince Henry the Navigator at Sagres, in Portugal. He gathered researchers there, collected maps and navigating instruments, and inspired others to sail down the west coast of Africa. He himself never ventured far, and commerce was not in his mind. But spreading the gospel was, and with it the true researcher's creed, which he expressed in the motto *Talent de bien faire.*†

Henry's generosity was not soon imitated: Sebastian Cabot got

---

* Bacon foretold the spread of research, American style, in the last pages of his utopia, *The New Atlantis.* After describing the several institutes and laboratories, he mentions the men: "We have twelve that sail into foreign countries . . . who bring us the books and abstracts, and patterns of experiments of all other parts. These we call Merchants of Light. We have three that collect the experiments which are in all books. These we call Depredators. We have three that collect the experiments of all mechanical arts, and also of liberal sciences, and also of practices which are not brought into arts. These we call Mystery Men. We have three that try new experiments, such as themselves think good. These we call Pioneers or Miners. We have three that draw the experiments of the former four into titles and tables. . . . These we call Compilers."

† "Only do a good job!" *Talent* here should be read in the old sense of will, determination.

no more from his English king for discovering Labrador than Milton from his publisher for the copyright of *Paradise Lost*; and it was not until the age of Enlightenment that governments again subsidized research expeditions. The aim then was to collect fauna and flora, discover islands (including Australasia), and improve ethnology by studying primitive tribes. In the nineteenth century, Huxley's voyage on the *Rattlesnake* and Darwin's on the *Beagle* were among the sequels to those efforts. Yet as late as the 1870s the British Association request to the government for a meager £120 to continue its study of tides was refused. This was long after Parliament had got used to spending thousands to add paintings to the National Gallery. Only by the turn of the century was science enough of a power in society to make research-at-large seem a proper investment. By then the British government readily gave tens of thousands to help measure the sun's distance by world-wide observations of the transit of Venus.

After another half century, today, England's zeal is second only to the American, and the mood in East and West can be discerned in the remark of a well-known physicist on returning from a trip to a Russian laboratory: "They're almost as overequipped as the Americans."[1] The will to "do research" and make sacrifices for it is so great that directing and refinancing are full-time jobs for hundreds of persons, and the planning and judging of projects produce shelf upon shelf of indispensable, indigestible literature.*

Research, in other words, is no longer simply a vocation; it is an institution. One is tempted to call it the chief symptomatic institution of our scientific society after the obvious one of powered machines.

A striking feature of Research is that in becoming an independent activity, born of science but no longer bound to it, the institution also broke loose from specified subjects. No one is surprised to learn that "Abstract Research Lights Up the Path for Businessmen"; or that we face another crisis: "While We Are Busily Hatch-

---

* To form an idea of the involvements, see President Pusey's Commencement Address to the Harvard Alumni Association, printed in the Association's *Bulletin*, July 7, 1962, 714, and Wallace R. Brode, formerly of the Bureau of Standards and now Science Adviser to the Secretary of State, "Development of a Science Policy," *Graduate Comment*, Wayne State University, vol. 3, no. 5, April 1960, 7-12.

ing the Golden Eggs That Science Laid Some Time Ago, We Are Letting the Productive Goose—Basic Research—Starve to Death."[2] No trace of irony should be seen in the comparison of research to a goose, nor of parody in an earnest message entitled "Research Explained" in the drama-page correspondence column: "The Research done for *A Country Scandal* was communication research. It was not 'motivation research' nor 'market research,' nor was it any sort of research done for invidious or underhanded purposes. The research was . . . to provide the director and producers with a picture of what was being communicated to the audience by the play in terms of mood, information and style."[3]

In keeping with its scope, the institution of research takes in many if not all kinds of persons. A researcher may be a chemist attached to a drug manufacturing firm or a girl who verifies facts for a weekly magazine; a canvasser who asks housewives what detergents they use or a strategist who designs a sequence of clerical operations for a business; a middleman who takes a fee for matching job descriptions with the listed capabilities of applicants in the file or a scholar attached to a university.

For many forms of research no special talent is required. Leaving aside the scholar and the scientist, we see why this hospitableness of research is welcome and how it answers the needs of a democratic techne. The mediocre abilities that enter white-collar employment desire the prestige of specialism; therefore occupations are made into professions and become academic subjects.* Practice is codified, textbooks are written, subspecialties multiply, and with them titles and degrees. Work and study fade into each other and are deemed equivalent. Young businessmen toil at Ph.D.'s in marketing, real estate operators write theses, personnel managers learn psychology to give tests, accountants, soldiers, diplomats, writers of advertising copy are taught in universities.

The many reasons for this latest transformation of trade by techne belong to different realms. Why, in the first place, has the

---

* Here too Bacon was a prophet. In *The Advancement of Learning* (Book II, sec. 23, par. 4) he advocates the teaching of business, by which he obviously means more than trade. No doubt he would rejoice at the rise of the practical arts we are now erecting into expertships.

impulse to action changed into a reflex of inquiry? Ask the manager of Cosmic Foods to open a branch in your neighborhood and he says, "We're researching." Talk of the menace of overpopulation, and the tycoon whose civic pastime is to promote world-wide birth control says: "What we need is lots more research." Bemoan the fate of the schools to the intelligent housewife and she retorts: "Can you wonder? There's a desperate shortage—" *not* "of teachers" but "of research in the teaching and learning process." Democracy in the sense of large numbers is here at work. Their preference must be consulted; their unsuspected differences must be found out; and what others have done in the matter elsewhere must be compared. All this is research, even when it is no more than a pooling of ignorance.

The march of science, moreover, produces the feeling that nobody in the past has ever done things right. Whether it is teaching or copulation, it has "problems" that "research" should solve by telling us just how, the best way. This alone is enough to stop the activist, and if those concerned happen to be a committee, one can be sure that nothing rash will occur for some time. The preliminary inquiry into how to begin the inquiry into the problem will be long and exhaustive. Approach is all.

Out of such research has grown a characteristic attitude and mode of speech which might be called pseudo-Socratism. It says with pride: "We can't know the answers, of course, but we're making progress about the right questions to ask." Or: "We didn't get very far, but it was a wonderful education for the committee." Or after writing up a two-hundred-page report: "We've only scratched the surface." The spirit of science, the being satisfied with the quest or game, is reproduced here, though its origin is social: the fear of being opinionated, overbearing, decisive, in short, the fear of unpopularity.

Besides, it is becoming ever harder in the mass world to judge situations intuitively and on the basis of experience, to say nothing of the difficulty of persuading others that one's judgment is right. And since after all action is often required, research mitigates the danger that small mistakes will be magnified by large numbers; a responsible manager wants to cover himself by "studies" and statistics—the findings with which he can placate his directors.

The private reasons for "researching" are equally plain: it is the shortest path to promotion and prestige; it frees the able young from drudgery and often entails paid travel. For an organization man to be "put on research" is the same as for a horse to be put out to grass—no stated hours or set duties. The very essence of research is that no precise result can be specified. The sense of privilege and of exceptional status that comes with research also obtains in the academic world to which it is native. It explains the popularity of honors work in colleges. It is a cause of the flight from teaching in universities—and of the many other flights to meetings and conferences across the world. The motive is not laziness or cynicism, any more than it is a passion for knowledge. Rather, it is a desire to escape the boredom of ordinary non-work. The flight is not so much from classroom study or from teaching as such, or from executive responsibility in a business or government office; it is from an excess of people, paper, confusion, conferring, and frustration.

Research affords the further pleasure of any specialism, that is, a comfortable monopoly, the emotional protection of "my subject": an adroit researcher manages to have a field as nearly as possible to himself. Society favoring an infinite division of labor, it is the mark of the second-rate to generalize, as it is of the depraved to popularize. But specialism need not mean solitary confinement. Small groups are happiest, and our adroit researcher enjoys the coziness of a subprofession, with set ways, a shorthand of speech and thought, and the friendship of his peers.* Such working arrangements provide in the midst of industrial life a kind of portable cloister.

Once on the right track, there is no risk, which is why the researcher is not guilty of ambition, at any rate not in the old sense. His effort is rather to define himself among the interchangeable parts of techne. This may be vanity, but it is not pride. By substi-

---

* The set ways are known as professional standards, which help to defend the subject from outsiders. Cf. Hayward's dictum that "the unoriginality of the professions is not always the result of excluding original men. It is an acquired quality." *Professionalism and Originality*, London, 1917, 85. For confirmation, see Roscoe Pound, *The Lawyer from Antiquity to Modern Times*, St. Paul, 1953, 354.

tuting for the amateur of British tradition the academic specialist of German origin, a scientific culture gets rid of much eccentricity and egotism, much impatience and unproductivity. The Rev. Dr. Chasuble was admired by Wilde's young heroine because, as she explained, "he has never written a book, so you can imagine how much he knows." Both defects are cured by our system, which infallibly turns every researcher into an author.

The outcome is already visible—a mandarin system which has organized its monetary support without destroying incentives. Research and its rewards supply no sinecures; they have to be daily earned anew, as Faust said of freedom: before a grant is a quarter spent, its renewal must be thought on.* Dr. Horace Miner, a close student of the subject who has codified its rules, informs us that "researching involves a well-ordered series of operations, the initial step of which is to secure the necessary funds to cover operating, living, and traveling expenses. The placing of priority on financing does not deny that back of every research project there must be an idea, but ideas should not be allowed to retard researching." And he reminds us that "the development of research ideas formerly consumed a great deal of valuable research time."[4]

The irony of these observations should not blind us to their accuracy. A ripe historian, who wants to enliven his subject by new kinds of research and who is not in the least ironic, expresses the hope that scientists will "endeavour to understand the true nature of history"; for as he truly says, "Scientists have the ear of those who hold the strings of the public purse."[5]

"Research support" always goes to novelty, the expectation of new knowledge. It is often called by the givers "venture capital for social progress" and it is first and foremost support of a proposal, only indirectly of men. Nothing is harder than to find money for scholars, teachers, books, and mere students, or for the fabric or the regular work of seats of learning. I am not speaking now of market research or industrial research, but of high disinterested research.

---

* The phases of the cycle are: proposal, project, "ongoing research," progress report, renewal-of-grant application. This being so, the project director is only a part-time researcher; he owes it to his staff to be an administrator.

On this score, our ways are both better and worse than those of the only comparable institution, the science and scholarship of Alexandria, which flourished from the third century B.C. to the second A.D. There the palace, which we may liken to the powers of government and business with us, maintained the Mouseion, where the arts and sciences flourished. The patronage was so direct that the staff poets and scientists had to be courtiers in their results as well as their conduct. Accordingly science prospered but art did not.

Our sources of support being scattered, money is less dictatorial, but it exerts a pressure even so.* It reshapes projects, encourages fashions, and generally puts greater value on procedures and accessories than on aims and ideas. Democratic gregariousness tends the same way. Projects have special merit when they require a team or a new togetherness (known as interdisciplinary) which is to link hitherto separate persons or subjects.† Projects are attractive when they call for traveling to other places, where men engaged in similar work may be interrupted with questions. Or again, projects should kindle a futurist sense of growth, perhaps by flowering into an institute or center. At the very least, the research should delve into the records or lives or opinions of many people and by its results promise to serve many more. And since the essence of research is inquiry, there should be interviews, questionnaires, or better still, a conference. So frequent and arduous is this enterprise that a handbook has been written for the guidance of researchers

---

* The mere amount to be had in recent years is in itself a power. Whereas in 1940 the total spent on university research was forty million dollars, in 1962 the sum was six hundred and thirteen million. Figures for business and industry are not clear cut, but the increase must be computed at an even higher rate.

† Teamwork and togetherness in Russian research is demanded on different grounds and with different criteria. "As a rule, it is not individuals but collectives that are able to solve fairly complex problems in modern science. The graduate [student] system, however, generates a tendency to individualism. . . . The country has an extensive network of research institutes in which personnel are obliged to do research. How do they differ from the graduate students who have also been specially selected for scientific research? Workers in a research institute have a compulsory working day, whereas graduate students are allowed free time for their scientific work and often do nothing for weeks on end. . . . It seems to me that the full-time graduate system should be eliminated." A student in Mathematics, Moscow University, writing in *Pravda* and quoted in *The Current Digest of the Soviet Press*, Jan. 14, 1959, 29-30.

faced with the care and feeding of other researchers en masse.*

The aim of research togetherness is an undefined "stimulation" by the transfer of ideas, persons, and "materials" from place to place. What this means is not that moribund researchers need to be galvanized by "contacts," but rather that the institution of research, like all others in a democracy, must justify itself—indeed, "sell" itself—to its constituents. The scientific principle of full disclosure to all judges reinforces the desire to achieve consensus, but the broader political urge is the stronger, as is shown by the vagueness and generosity of the conditions imposed on research.†

So much for the externals. Within the cult of research lives an old faith to which justice cannot be done in generalities. One must have seen, as it has been my lifelong privilege to see, the calm energy and light-bearing intellect which in the best universities produce the great work still being done in science, history, and the other philosophic disciplines. Unfortunately, the atmosphere of the more worldly cult—a sort of upper-mass culture—is not favorable to the true scholars. When science departments were established in universities seventy years ago, other subjects gave themselves over to a professionalism which is now unshakable. Making the Ph.D. degree an admission ticket to teaching and re-

* Lucien R. Duchesne, *Congress Organizers' Manual*, International Congress Science Series, vol. 2, Union of International Associations, Brussels, 1961, 98 pp. and charts. An idea of the work may be had from one or two quotations: "Supposing the optimal conditions to be satisfied by one or more towns, the OB [Organizing Body] has to make its decision" (10). "When H hour and M minute arrive, all the pieces must be in position on the chessboard so that all moves can be carried out faultlessly" (14).

† A few years ago, the American Council of Learned Societies held a conference to discuss these conditions. It proved impossible to state clearly "how selection of those supported should be made, though there was considerable support for the argument that the quality of the questions the applicant is asking, and the degree of precision and the level of sophistication he brings to the study on which he means to embark should be prime bases for decision. Publication records . . . are also important indexes.

"It was also generally agreed to that the nature of the project itself must be considered, and a great deal of time was spent in discussing the criteria to be used in judging significance, with recognition that only the most general criteria could be applied to all the fields represented. . . . Some of those present thought that social need should be a basis for judgment, while others championed 'pure' scholarship, which might not have social relevance." *A.C.L.S. Newsletter*, vol. XI, no. 5, May 1960, 3.

search let in the seven devils of careerism; research money did the rest.*

The desire of all scholars to emulate physical science and of all universities to house none but scholars has been dearly paid for. Original work, a "contribution to knowledge" is required where there is neither talent, impulse, nor matter to make it out of. Some disciplines have run out of suitable subjects: apparently all is known. The truth is what the head of a distinguished department of English on the west coast once confessed: "If research is the right name for good work in Science, then it is the wrong name for academic scholarship, good or bad."

To illustrate from the literary studies he had in mind, the topics treated in doctoral dissertations or professorial articles often distort the art they discuss by abstracting elements of perception or taste and turning them into things, in the manner of science. Thus someone will "examine the later imagery of Tennyson." The ultimate aim of the work is to be cited in somebody else's work as a scientific "has-shown"—"Mr. Jones has shown that the later imagery of Tennyson differs markedly (or slightly) from his middle imagery." There are obviously three articles—early, middle, late—to be made out of poor Tennyson's melodious metaphors. Such scholarship reflects the influence of "method" in criticism. The logic goes: imagery is a product; products can be sorted, counted, X-rayed. So Tennyson's mind is to be treated like his intestines after a barium meal; which gives later scholars the chance

---

* The co-inventor of the American Ph.D., John W. Burgess, foresaw the evil as well as the great democratic demand for titles and labels. See his report as Dean of the Graduate Faculties, Columbia University, for 1912. And for contrast, the statement, half a century earlier, by the physicist and college president, F. A. P. Barnard: "Our scholars are self-made. Their scholarship is the growth of their mature life." "Early Mental Training, an Address to the Convocation of 1866," reprinted in E. L. Youmans, *The Culture Demanded by Modern Life*, Akron, 1867, 315*n.*

See also the remarkable "Letter to a New Ph.D." by William G. Carleton in *Teachers College Record*, December 1961, 199-207: "What a change has come over the academic world since I entered it over thirty-five years ago! How quiet, unworldly, and innocent it was then, how modest the material surroundings and rewards, how few the opportunities for glamorous careers. . . . Remember, Rip, you are now entering what has become one of the bitchiest professions in the world. Don't let your guard down!"

to show that the previous has-showns were quite wrong.*

The further expectation—still based on analogy with science—is that the little articles and the overlong books will someday come to-gether to form a Great Scheme of knowledge. Yet scholarly cau-tion often makes a man re-do the work in all the earlier has-showns, and synthesis is rare as well as risky. Few would subscribe to the definition E. M. Forster once gave of the scholar—a "man who chooses a worthy subject and masters all its facts and the leading facts of the subjects neighboring";[6] for specialism kills the judg-ment with which to distinguish a worthy subject, to say nothing of the self-confidence needed for large ventures.†

Instead, the mass output of small "contributions" results in the great paradox of abundance: our culture is enormously productive of knowledge as of goods. We know—as the saying goes—more than ever before. But who is "we"? And *where* in a practical sense is the knowledge that our many independent groups of specialists produce?

Scholarly research could plead in extenuation that its wasteful-ness is no greater than the scientific, while it lacks some of the scien-tists' advantages. The historical disciplines have no laboratory in which hypotheses can visibly go astray, blow up, give out an un-mistakable negative. There is not even the daily exchange of views with colleagues over a tangible piece of apparatus: all these guide-posts and retaining walls are absent in book work. But the cult of research disregards such differences. It decrees that universities exist to increase knowledge and only incidentally to impart it. The increase must be palpable; hence every academic man shall publish learned work. Salary and promotion depend upon publication. A scholar who husbands his time and his notes can make a small paper out of every unexpected fact he finds as he progresses through a

---

* The scientific pattern can also be imitated in scholarship by professing to study "X *as* —" something else, again in the manner of science. That excellent scholar (old style) Mr. Gordon Ray pointed this out after reading three inde-pendent essays that sought to interpret George Eliot's *Mill on the Floss* as water symbolism. "What they had in common was a profound misunderstanding of the nature of the Victorian novel." "The Undoctored Incident," Address at New York University, April 21, 1961, 12.

† In prefacing his history of English literature (1960) Mr. David Daiches made apology and begged indulgence for being only one man and not a team of specialists.

larger project. The journals in any field are accordingly two or three years behind the profession's output.

When the scholar's large work is finished, it is very likely difficult to publish. Often it is unreadable—some trade publishers say "illiterate"—and usually it is too long. Research grants seldom provide for publishing the results, the foundations taking the view that merit should justify publication as a trade or university-press book. In fact a good book, as defined by the scholarly community, will sell from a thousand to fifteen hundred copies in ten years, a fifth of the sales being to libraries.*

The large reading public shows its respect for academic scholarship by keeping its hands off, and scholars will not buy what they can find in libraries. Publishing, therefore, chiefly serves the end of professional competition. The judgment of a handful determines rank within each specialty, but that rating is all-important. Since the second world war, the desire of colleges and universities for prestige through research has become a fever, an obsession. The competition for the men, young and old, whose work is approved by the knowing is ruthless. They are offered high salaries, perquisites, and virtual freedom from teaching. The bidding institution hopes that the incumbent once bought will yield the rich butter of prestige, churned from the pure cream of research. After the best men have been sterilized as teachers—made into drones of research —such undergraduate courses as they might have taught can always be inexpensively handled by graduate students.

This competition has radically altered the traditional aim of universities as well as the substance of undergraduate programs. It is natural (and also required) that the young men who enter academic life should follow the practice of the profession. The doctoral thesis imitates the older scholar's book, often to the point of caricature. With few exceptions these dissertations are ill-organized, poorly written compilations of what is already in other books. Everybody agrees that they take much too long to prepare and that with greater speed they might undergo a desirable Fitzgerald contraction. All are microfilmed, catalogued, and stored, and some

* See Dr. Rush Welter's instructive *Problems of Scholarly Publishing*, American Council of Learned Societies, New York, 1959, 67 and *passim*. No encouragement is given to short books because the lower price that must be put upon them will not cover the expense of production.

good ones are published; but their contents are not in much demand. The upshot is that the real test of the young scholar is his first work after the dissertation, and since this is difficult to produce while carrying a full teaching load, provision is made for leaves of absence. This is but fair if "a good book" is absolutely required for advancement. The system has thus unwittingly established the Right to Research.

The combined result of the struggle for research time, research men, and research money is the distortion of the college curriculum. Having persuaded themselves that only research is important, undergraduate instructors shape their courses to be introductions to their several professions. Every sophomore who takes Anthropology I is treated as a potential anthropologist; every senior who elects Milton or Shakespeare is taught as if his aim were to edit texts. As in the physical sciences, the assumption is that there is "nothing to talk about" except the techniques of doing—doing research.

Long familiar with the research ideal, many students respond. The best elect honors work, which is precocious scholarship, "independent research." That frees both student and instructor from the yoke of regular class hours. The taste is formed: in the United States, nearly half the graduates of the leading colleges take a year or more of graduate work with no intention of becoming scholars. They know that doing research will help to impress the future employer. He will see that the candidate has been "qualifying" for him from early days and can be counted on to appreciate, as a junior executive, the "values of research."

This prospect brings us back to the worldly uses of research: they follow academic forms, not only for ostentation, but also because the producers are academically trained. The report, with its background opening, its definitions, and its statement of assumptions; the footnote and listing of references; the tables and graphs, themselves borrowed by the scholar from the scientist—all come naturally to the business researcher, who employs them for himself and also as elements of persuasion in advertising. The look of research engenders trust and allays the anxiety of choosing among the goods of techne.

At times, research shows that millions favor a product. At other times, research ascertains what the millions might like if they were offered some object they have never seen—it is a survey of ignorance. Again, research "tests" the existing article,* and the reader feels the drama of science—the heroic struggle with nature yielding at last the secret and the power. While this goes on in the firm's laboratory, in another department the counterpart of social science is going on, for after producing the cigarette which will appeal to all smokers, it is instructive to learn why all do not smoke it.

So much "research" is published, and the ulterior motive is usually so clear, that the public rations its attention. But something remains, for the ritual appeals to one of our chief pieties. In a scientific culture, research is the universal solvent of problems, and in a democratic culture it affords the further satisfaction of a worthy self-consciousness: we "study" ourselves, we let others "study" us —all is aboveboard and the results are numerical and definitive like an election.

These familiar doings make the honored name revert in common speech to its simplest meaning: research means "finding out." People in doubt over a spelling will say that they "did a little research in the dictionary," just as the advertiser lets them believe that they are "experimenting" with a new toothpaste. The citizen, in fact, began his research career long before college or business school taught him professional methods: as a school child he was asked to collect travel folders and with scissors and paste make them into a "research report" on a foreign country.

If the influence of the cult stopped with these verbalisms and sentiments, one could limit one's worry to what is happening in the universities. But the damaging effect goes deeper. Dr. Miner puts this tersely in the title of the essay I quoted from: "Researchmanship: the Feedback of Expertise."†⁷ The meaning of this is that

---

* "Credit for Kent's Sales Leadership Goes to Research. . . . The success of Kent [was credited] to the P. Lorillard Company research laboratories in Greensboro, North Carolina. For many years, Lorillard research scientists have been experimenting in order to create a cigarette of such excellent taste quality that it would appeal to all smokers, yet with a lower tar and nicotine content than all other leading brands. In 1957, the years of research were crowned by the development of the new Kent."

† See above, p. 126.

the rise of research from a talent and special training to a common pastime and a mode of rhetoric changes the national intellect and character.

A first sign of change is what I may call the habit of Pseudo-Analysis. Its model is the opening section of a genuine scientific report which recites the nature of the problem. In the imitation there may be no problem at all, or it is one of which the "analysis" offered is platitude. For example, in the manual previously cited on the organizing of conferences, we read: "The period, length, and frequency of the congress affects its organization; similarly with the type and number of its participants."[8]

At other times an abstraction is dogmatically split into several others, though it is clear that the subdivisions could be half as many or twice as many and could bear different names without altering one's understanding of the subject.† In this type of analysis, beware especially of processes. Whenever one reads, "Let us first take a look at the basic elements of—" the democratic process or the teaching process, or the process of communication, or of city planning, or calisthenics, one is almost sure that what follows is filler, and usually tautology.

In scholarship other than scientific, "analysis" accounts for much of the unbearable length. The judgment of what is good work has changed and now includes as a matter of course the claim of completeness, an artificial orderliness (which leads to intolerable repetition), and the practice I spoke of earlier apropos of Tennyson, of making matters of taste or insight into abstractions amounting to natural processes: not "Dickens' frequent failure to create true heroines," but "his psychologically rooted incapacity to let his imagination play with sufficient freedom in the creation of lovable women." One might be listening to the diagnosis of a malfunctioning liver, except that the liver is a real organ and its "processes" are fairly well known. The pretentious analyzing of every subject in this fashion is what has made reading laborious for small returns.

Pseudo-analysis inevitably adds to the sum of jargon and fosters

---

† On "the decisionmaking [*sic*] process": " 'Harold Lasswell refines the process for purposes of functional analysis into seven steps: intelligence; recommendation; prescription; invocation; application; appraisal; and termination.' " See Lasswell, *The Decision Process* (1956); Donnelly, Goldstein, & Schwartz, *Criminal Law,* Glencoe, 1963, 1*n*.

the replacement of explanation by metaphor, which I shall later discuss. Behind both is the desire to "formulate," as in science, and indeed as in advertising and techne—all of them pursuing the invisible yet picturesque process behind the product.

Also born of research is the love of excess in supplying "background"—the history and antecedents of every project, idea, or situation. This excess, if challenged, would be justified by an appeal to the all-importance of fact, which slips easily into the importance of all facts. In journalism, government, business, popular literature, conversation, and high art, the use of fact to rout indifference or disbelief is preferred to the use of ideas. With the abundant sources at hand it is easier to write a factual than a philosophical book. The popular work of travel or biography delivers a barrage of small and justly neglected facts—about times, places, distances—like the so-called profiles of living men, composed not of significant lines but of factual dots. The principal form of fiction, the novel, is of course factual to the limit where the glut of fact destroys coherence, as in the fad Dashiell Hammett started of inventorying the clothes, food, drinks, and ablutions of his characters. And only in a factual society would it be necessary for writers of palpably made-up narratives to state that the persons and places written about are imaginary.

The paradoxical failure of many fact-finding reports to help the public understand its own difficulties is almost always the result of Factuality. The committee appointed to "study the facts" is itself enlightened by them, but in publishing too much of what it finds and not enough of what it thinks, it leaves the interested citizen discouraged and the large tomes untouchable. Abundance here also brings no pleasure. Specialism is not to blame except as it feeds the lust for fact, which in turn is a cause of boredom. The connection has been noticed by outside observers as a peculiarly American evil.* It is now growing wherever techne, democracy, and research extend.

Under this regime, every group, however temporary, keeps

---

* Tocqueville noted it in passing; Dickens was more outspoken; and the gentle G. Lowes Dickinson was driven by pain to cry out: "They bore me as it is impossible to be bored in Europe, and that is because *any* fact interests them, and *no* idea except as it can be shown to be in direct relation to fact." E. M. Forster, *Goldsworthy Lowes Dickinson*, New York, 1934, 131.

minutes, issues reports, and publishes a newsletter; it is ready, if necessary, to stop the activity for which it came into being so that the record may be "kept straight" and "made available." Only the criminal classes hold to the ancient reticence and wish the facts to stay unknown. The citizen-researcher would be shocked if he were told that he may be as vulgarly greedy in amassing facts as the tycoon in amassing dollars. Yet as John Stuart Mill—no enemy to science—foretold, the effect of factuality is a paralyzing clutter, a fatal decline of intellect: "What will soon be the worth of a man, for any human purpose, except for his own infinitesimal fraction of human wants and requirements? His state will be worse than that of simple ignorance."[9] In other words, the appeal to fact, from which so much good was expected and did result during the hopeful time of enlightenment, is now turning into a blanketing darkness.

The presumption of a technical civilization that all good things can be provided meets in this case an instructive denial. When he was self-appointed, unstimulated, perhaps unrewarded, the searcher after truth possessed an impetus to do the work, which might imply talent and perseverance. When, on the contrary, search turns into the subsidized routine of research, all that is looked for and most of what is supplied is a steady income of facts. The ability and the will to interpret diminish.

A scientist attached to the National Science Foundation attributes this enfeeblement to overspecializing. He calls it "the precipitous hazard of undergeneralizing."[10] But since intelligence is as frequent in the population as before, what keeps the trained minds from the desired generalizing? The answer, I think, lies in the repression which the love of fact develops, the fear of Error as absolute evil. That many errors are negligible and should not occupy the mind is forgotten in the ritual of accuracy.

In this ritual fear is proportional to visibility: a date, a middle initial are incontrovertibly right or wrong, and the proof of error is painful. But the effort to be safe first misdirects attention, then corrupts judgment.* This is illustrated by the veri-

* Note the curious balancing of merit and demerit in a comment by a young scholar upon one of our ablest (and most accurate) journalists: "Although I gen-

fying that goes on in editorial and publishing offices. Trivialities are set right and large errors of logic and sense pass by unseen. This is deplorable even in journalism; at every higher stage of intellect the bias spoils more and more important elements. As Judge Cardozo observed many years ago, "There is an accuracy that defeats itself by the overemphasis of details. I often say that one must permit oneself, and that quite advisedly and deliberately, a certain margin of misstatement."[11]

In truth, the commandments derived from science and techne and clustering in the institution of research boil down to a diffuse, pervasive fear—fear of fellow professionals, fear of one's own mind and ego as "subjective," fear of errors creeping into a text while one's back is turned. The resulting temper is not, on the face of it, favorable to good work. Indeed, what fear inspires is work other than that which might spontaneously attract the mind; the job must fit the means and curiosity be redirected to questions ("problems") that can be managed without running the prohibited risks. The mind, in short, is subdued by the mechanical.

Mechanics may mean lists and facts and "theme analysis" done by hand count; or it may mean reliance on the innumerable little studies produced under the prevailing scheme. Thus a leading Shakespeare scholar in this country states that he has time and strength to be "only a bibliographer"—the *substance* of scholarship is held in escrow for a later people. Or again, mechanics in research may be the preference for visible facts, usually small, on which large inferences can supposedly rest. To stay with Shakespeare another moment, some scholars believe the study of his works has been revolutionized by the examination of the type and the way in which his plays were set up in early editions. From the type the typesetter is inferred, from him his characteristic mistakes, and from those the true text. Compositors A and B are now figures as familiar as the five majuscules who put together Genesis, and a man's reputation in research was made when he "brilliantly identified apprentice compositor E."[12]

---

erally find James Reston perceptive and stimulating, he incorrectly states that only twice in the last one hundred years, etc." *New York Times*, Aug. 4, 1962.

Inferential knowledge is peculiarly attractive to the men of a scientific culture; it seems more knowing than direct information; it has the charm of detection.* And this in turn invites to studying everything *through* something else: the truth always lives next door.† For this, mechanics in the literal sense is now available. Electronic computers, which are plainly useful in physical and social science, are being toyed with by humanists in the hope of demonstrating their truths by number too. "Proofs" have been given of this or that hypothesis of authorship (Homer, the Dead Sea Scrolls, Nathaniel Hawthorne); just as in the studies of painting, infra-red photography has supplied ludic man with aesthetic inferences which should perhaps be taken with a grain of developer.

In these gropings the root assumption obviously is that the subject of study—man or one of his works—was produced mechanically, like the instrument designed to reconstruct its meaning or genesis. In other words, men are machines, whether as authors or painters, voting citizens or typesetters of plays. And they run true to form; for we cannot with a machine allow for accident, self-contradiction, perversity, laziness, or folly.‡ Of these traits we rule out, we have, needless to add, long since forgotten the *utility*.

And yet, the observer of the strange scene is surprised to note that the practitioners of mechanized, error-free research have one word continually on their lips. That word is "Creativity." As a claim, a hope, a promise, the label "creative" is nowadays attached to whatever is good in the life of man under techne. "In jazz and folk singing we want programs that have a creative point of view."

* The art of detection was first appreciated in the Enlightenment, as the skits by Voltaire and Beaumarchais testify.

† A famous example of this occurred a century ago, when discrepancies in Scripture were energetically studied. Soon nothing was to be accepted as fact on biblical authority because it was *written*. What the text was thought to need was confirmation from archaeological evidence. Meanwhile, the archaeologists were studying civilizations through imperfect remains, of which they would have gladly bartered a few shards for a scrap of written document.

‡ Awareness of this premise is implied in the hoax devised some months ago at an electronics convention in the west. "Linotron" was described as "A Practical Device for Majority Logic" and it was meant to answer the question "Can a foolproof method be devised for selecting valid, high-quality technical papers for major technical conventions?" *Chemical and Engineering News*, Aug. 6, 1962.

"Mr. ——, recently appointed head of the studio, would make the creative decisions." As regards "The Greatest Problem in Marriage—Communication," a young couple are advised that "if they look out and see it is raining and decide to stay home and have fun reading, they have creativity, the highest state of good marriage adjustment." "A teacher can help children develop creative thinking and behavior in arithmetic."[13]

If these uses of the word and the idea seem vitiated by their journalistic source, we can with little effort find the same intention in more considered writings. Thus Sir Herbert Read tells us that we should work for "illumination," a "light that is released by the creative energies latent within every child and every adult."[14] And he adds that this idea "is necessary to balance the prevailing creed that knowledge is power." Creativity, clearly, is the opposite of the mechanical, it is anti-techne par excellence. It stands for the desire to make something expressive of the whole being, something private and joyful, fashioned in untrammeled ways; something which moreover need not be justified to others, because it is self-justifying.

The users of the word are very likely reaching back toward the old idea of Work. It is when speaking of work that reference to creativity most often occurs. The employer promises creativity as part of the perquisites: he says that the job will be "exciting," "essential," and that it is up to the incumbent to make it "truly creative."* In giving an account of one's work to others, contentment or (more often) the effort to put up a good front will alike be expressed by the term "creative"; and when openly dissatisfied: "that place gives a man no chance to be creative."

This pathetic use of a once awe-inspiring idea helps to explain the paradox of freedom in work—does it lie in "doing research"? Yes, for it means escape from routine tasks. But again, No, because research limits like all specialisms and is subject to the dull conformities of the fashions and the professions. When creativity is discussed, as it often is in academic and business conferences, the call is always for Imagination; that power is regarded as the source of creativity, and one quickly sees that what everybody wants is

* A series of ads a few years ago bore the caption: "The Creative Man—Don't Fence Him In!"

the blessedness of being a poet. The advertiser and the employee, the married pair and the art student want nothing less than to combine in their lives the joys and the glory of the only honored vocations. They want to escape every form of non-work. They would gladly be known as artists, if at the same time they could claim the proficiency and prestige of scientists. The executive-cum-researcher envies these two sainthoods equally. Having neither he wants both. Caught in the single managerial routine he makes a try for freedom, first through research, and when this turns out inadequate, through an additive—creativity. That is why there are not two cults but one—Research-and-Creativity.

The grave objection to the worship of the Janus-faced deity is that like most modern cults it is not sober, and soberness is essential to ordinary men's enjoyment of genuine, plain work. The work of genius in art or science is something else. A great artist or scientist may think and talk nonsense—it rarely interferes with his work. But mere able men are spoiled for theirs by the kind of nonsense that leaves geniuses unharmed. They are unharmed because they do not believe in it with the simplicity of the man who, for instance, drafts a new "program of creativity" in his business or in a school. If he knew what creativity really was he could not require it of his men or his boys. Ours is the first time in history that a man has been *expected* to be creative—and by means of *research*!

The natural and not invidious antithesis between creative and academic has thus been obliterated, and colleges and universities unblushingly offer courses in Creative Writing. Students take them, if only because the name suggests freedom from whatever is regular. What is the result?—another opportunity for avoiding the discipline of words and the responsibilities of feeling and thought. "Creative" writing will express "individuality" instead of common form.*

* As against the college definition of creative, one may compare the casual remark uttered by a truly creative writer long before he made his name. D. H. Lawrence, aged 27, confides to his former fellow-teacher McLeod: "I do very little work—just the revising of this novel—no creative work." (*The Letters of D. H. Lawrence*, ed. Diana Trilling, New York, 1959, 14.) If Lawrence revising *The Trespasser* is not creating, what is creative about the undergraduates? Can it be that they don't revise?

But to denote the educational value of some free writing by the name of creation is to mislead the young, while destroying the work done elsewhere on the campus by the Departments of English and Comparative Literature, of Classics, Fine Arts, Music, and History: they try to explain what it was that Milton, Homer, and Hugo, and Mozart, Gibbon, and Michelangelo achieved with the common materials available to all men. That is properly creation. It is difficult to do and difficult to understand. And it follows that the innocent aping of current work, whether commercial or highbrow, bears no relation to the thing we dignify by a metaphor that compares it with the act of a god.

For if the useful and worthy talents of the market place are creative, then everybody can have his democratic share of fame; if research consists in publishing papers and nothing more, then we are all safe. In the name of creativity, thou shalt not damn my research and I will not even look at yours. Our brethren will also respect this compact—for a price. They will think little of what we have done—and we shall think little of ourselves. Meanwhile, to cry "Hands off! Creativity is going on!" is the right of every man, just as research is the duty of every *academic* man. And to make every man both academic and creative seems to be the manifest goal of evolution.

In the process, the tremendous range of human powers contracts. When the child in the kindergarten is called creative for his first finger painting, the distance between him and Rembrandt is officially shortened. "Potential" is there; the futurist society is content to wait. And for the adult who is not Rembrandt, creativity fills the void of passion that is inevitable when one works in answer to vague "pressures" rather than to supply the visible needs of men. Seeing this, one must also see that creativity cannot be a *goal*, in education or anywhere else. *Search* cannot be standardized. Professional research is not necessarily pedestrian, but the ability to "do research" is not to be confused with a capacity for work.

If anyone doubts this, owing to the force of the egalitarian cult of creativity, one need only ask whether the sight of "research" as it now flourishes ever "takes your breath away, or sends a momentary wave of coldness across your face, or elicits whatever your special bodily signal may be of your mind's amazed and sudden surrender to some stroke of passionate genius."[15] To ask the question is almost indecent, but it is necessary; and this not only to close an unhappy chapter on the modern ego, but to open another on the ways in which science can and cannot be made the mental property of everyman in a scientific culture.

VII

Science the Unteachable

IN recommending soberness to those who think they can provide for creativity at will, I have not meant to belittle the strength and worth of the desire that is eating at their hearts. On the contrary: theirs is the very passion that I see balked by techne and, because it is unsatisfied, killing man from within. But I also see that the make-believe or verbal satisfactions now offered are not going to appease the yearning. All that the substitutes can do is to weaken judgment, and hence make still harder the path of the creator, postpone still longer than in the past the acceptance of true creation.

When creation is spoken of as being within the power of man, it is most often linked with the work of the artist, and in that form it is rightly felt to carry part of its nourishing virtue to the beholder. This being so, there is cause for concern in the fact that although the western world has witnessed during the past thirty years a veritable art-rush, the effect has not been general fulfillment and joy—rather the opposite. I shall try later to suggest why this has happened. Meanwhile I must deal with that other belief of our scientific culture about creation, which is that it characterizes also the work of the scientist.

I pointed out earlier the important difference that limits the force of scientific creation upon the beholders—very few are able to feel it and those few can only receive a very abstract pleasure. Sensuousness and passion are necessarily absent. But this may deprive only the beholders. The makers of science can experience all the passions of a creator and their senses doubtless respond like the artist's to their familiar materials, especially when the science requires working with substance rather than its traces.* There is of course one further limitation: the scientists who can be said to perform the creative act are even fewer than their fit beholders, fewer also than the creators of styles in art. The minds of that small company are as solitary and strange as Wordsworth said in his eulogy of Newton, and the gap between them and the crowd of research men is expressed by all the distance that separates creation from making by method.†

What is true of the great artist—that he cannot be taught because he is destined to teach others—applies even more strictly to the world's few scientific geniuses, precisely because of that strangeness of mind which enables them to imagine or see into the scheme of *things*. The more ordinary yet still uncommon talents that fill the ranks of the scientific profession are, we know, taught as future professionals almost from the time that their bent declares itself, and it is probably fair to say that in the leading western countries the full-fledged scientific worker is as well trained as anyone has ever been in any other guild: more time, care, expense, equipment, and tests of knowledge are put into the preparation of these men and women than the world has ever bestowed on its soldiers, teachers, lawyers, or clergy.

And yet those in charge of instruction are dissatisfied. The schools of science, engineering, and medicine complain of the poor training received in the grades below. The leaders of the profession worry about falling enrollments in scientific subjects and are perplexed by the transfers to other fields of those that set out in adolescence to become scientists. In a culture that rewards scientific talent as generously as any other work of the mind, the situation is indeed astonishing. Special inducements and propaganda to create an atmosphere favorable to science as a profession help recruitment only a little.‡ Among the young, the knowledge even of what the well-

---

* There are marked responses to *types* of material among scientific creators too. Newton was best when working at subjects that taxed his deductive powers; Faraday was careless in theorizing but had a flair for the "go" of concrete experience.

† "Crowd" is not an invidious word when one remembers that ninety per cent of all scientists who have ever lived are alive today. From that mass it would be hard to choose with reasonable certainty a Galileo, Pascal, or Newton.

‡ Anyone interested in the history of the effort should look up the Symposium published in *The Scientific Monthly*, vol. 82, no. 6, June 1956, and go on to the

established sciences deal with is scarce, and the heavily advertised national need for the technically trained produces no sufficient supply.†

Since the postwar generations of college students have been anything but self-seeking, hoping rather to "serve" in an absolute sense, the failure of science to attract them has led to the suspicion that the fault lay in the curriculum. Accordingly, with characteristic energy, school boards, foundations, teachers, and experts in education have surveyed the difficulty and set up committees to remedy it. Here the answer was "more science hours"; there it was more money for equipment and assistants; elsewhere it was "crash programs" under high school teachers who had been retreaded (a technical term) for the purpose; and throughout the nation senior scientists of repute gave up time and effort to revising radically the presentation of their subjects.

No sooner were these innovations under way than they were attacked. Dispute filled the air about the excessive ease or difficulty, the efficiency or "objectives," of the programs. The public were bewildered with claims and figures; and the teachers with new texts and unaccustomed expense money.*

In England, the same issue of More Science continues to be argued, though with greater consecutiveness and a stronger critical spirit. It was in favor of more science that the University of Oxford decided to drop Latin as an entrance requirement in February 1960 and to establish a new Honors School of Engineering Science three years later. But England is divided on the question whether the training should be for technicians to serve industry or pure

---

numerous editorials and articles published in *Science* from January 1958; then study the descriptions of the National Science Foundation Programs—Undergraduate Science Education; Summer Science Training Program for Secondary School Students; Cooperative College-School Science Program; Research Participation for Secondary School Teachers; Research Participation for College Teachers; finally, the last four chapters of *The Scientific Revolution*, ed. G. W. Elbers and P. Duncan, Washington, 1959.

† The demand is often stated in Utopian-futurist terms. As Mr. Derek Price points out in *Little Science, Big Science* (New York, 1963), estimates based on the present rate of growth would require that in one hundred years there should be two scientists for every man, woman, and child.

* In speaking at a summer institute for high school teachers of physics, I found that increased appropriations for science often went to waste for lack of space to house new equipment and lack of need for supplies.

scientists to pursue research. The second choice, which entails the maintenance of large university departments, has been denounced as a "vast Americanization" that would result in pure waste.[1]

The current state of affairs in the United States can be inferred from the report of Montgomery County, Maryland, which admits that it is easier to discuss an ideal science program than to set it up. The chairman of the committee was not even sure that they had "decided what they were trying to do."[2] Is it not, once again, astonishing that a scientific culture that has the means, that can draw on expert advice, and is secure in the tradition of a science requirement, finds itself at a loss, not alone in carrying out, but even in thinking about the teaching of science? One great deficiency has been of course the inability of parents and others on the lay boards of schools and colleges to express an opinion. Of all subjects, science was certainly the last the public would have supposed needed their pedagogical guidance. And when they heard of the need they were helpless from knowing little and fearing much. Their predicament and that of the nation make one wonder whether science is in the usual sense teachable.

The one undebated idea has been that of the science requirement. It escapes notice by being familiar and perhaps also by being futile. For, obviously, its intention in school and college is double: to "teach science" to everybody and to attract and begin to prepare professionals. The recent outcry seems to show that it does neither, the scientists we do recruit being rather predestined than lured by teaching. At the same time, it is difficult to be sure what the intention of the requirement has done to the teaching of the subject, and hence one may usefully ask what the originators of the scheme had in mind.

Within the common curriculum that double intention is peculiar to science. There is no "law requirement" designed to educate citizens and recruit lawyers; and the other requirements—in English, history, foreign languages, and the like—are not meant (at least not explicitly) to supply poets, historians, and scholars in foreign tongues. Such men choose their calling from a public awareness of what it is to be a lawyer, historian, or businessman. Yet this special blankness about science did not always obtain among the educated. It dates—and this is one of the many paradoxes we keep stumbling on—from the re-establishment of science within the curriculum in the late nineteenth century. In the Middle Ages, when Plato's educational philosophy was revived, his requirement of geometry as the base of all learning prevailed.* The Renaissance and Reformation reacted against this in the name of the humanities and gave more time to the study of the literary classics, though mathematics remained without interruption a part of liberal schooling. It was this combined medieval-Renaissance tradition that the men who came after Darwin attacked in order to force science into the schools. They were not battling literature and the arts. They were attacking Latin and Greek, on the one hand, and logic and mathematics, on the other. They argued that the study of these four subjects overtrained memory and the deductive faculties. The lack of modern physics and Darwinian biology, they said, left observation and the inductive faculties undeveloped, though these neglected talents were strong in nearly every young child. Disuse let them atrophy in all but a few.

Campaigns in Scotland, England, France, Belgium, Germany, Italy, and the United States brought about the desired changes. In Germany the quarrel was about practicality. Huxley's dictum that industrial progress depended on making science common knowledge was influential but not decisive: rather, science came in as an "antidote to logic," as a corrective to the "monotony of reason," and of course as an offset to the remoteness of the classics, especially when taught as philology rather than as literature. In short, science was to be the cure for excessive abstraction.†

---

* In the *quadrivium*, or upper course, three of the four subjects were arithmetic, geometry, and astronomy. Over the gate of Plato's Grove of Academe were the words: "Let no one enter who is without geometry." This science requirement was new with him and his elder, Socrates. One generation earlier, Aristophanes could still make the Athenians laugh at the idea of teaching astronomy and other unpractical subjects. *Clouds*, 220 ff.

† For Huxley's pronouncement, see his famous lecture "Science and Culture" delivered at the opening of Mason Science College, Birmingham, in 1880; and for a wide survey of English and American opinion on the question, *The Culture Demanded by Modern Life*, ed. E. L. Youmans, Akron, 1867.

The new educationists approved: they were trained in German scientific psychology and they believed that the "natural order of learning" was from things to ideas. Mastering natural science would moreover endow the mind with "superior and exact common sense" and supply "what we need to know in the modern world."[3] In being worldly to this extent, the new curriculum did not so much oppose a liberal education as the *clerical* education that had ruled over the minds of clergy *and* laity for three centuries.

This sketch is enough to show how far we have drifted from the original purpose. Science is no longer natural observation and common sense opposed to logic and mathematics; it is itself abstract and mathematical; it is no antidote to the memorizing and the Latin which have long since ceased to count; it is itself imposed by the weight of tradition; and of those who bear its rule three quarters are still an unimproved laity.* One can say about science exactly what the physicist and college president F. A. P. Barnard said ninety years ago to persuade his colleagues that the burden of the classics was pointless for the young: "They do not lift it, though we may persuade ourselves that they do because we tie them to it and leave them there."[4] The science requirement is perhaps less of a burden because it starts later in school life than the dead languages used to; but in college science is wasted on all the future laity. It makes them no more "literate in science" than they used to be fluent in Greek.

What is equally bad is that the professional scientist also fails to be liberally educated in science. Like his predecessor the philologist, who killed the classics by pedantry, he is as to science a philologist new style. And as we saw apropos of research, the charm of the scientific stance is affecting instruction in other subjects: "It is frustrating for the scholar," says a university professor, "to be forced to impart second-hand knowledge to his students and be unable to teach that which is part of his on-going research life."[5]

I am not here concerned with the merits of this doctrine but with

* One item of this failure in the United States, as George Gaylord Simpson reminds us, is the refusal of many public schools to teach evolution, because of religious or anti-intellectual prejudice. "One Hundred Years Without Darwin Are Enough," *Teachers College Record*, May 1961, 617-26.

its influence, which helps to explain why the teaching of science does not answer expectations. To teach only one's own discoveries is, in an age of specialism, to make the student see the world through the eye of a needle. Making allowances for the exaggeration of emphasis, the truth is that, despite the pieties in journals and panels, the ideal of liberal education is dying. The institutions that continue to believe in it no longer find willing teachers. "The recent Ph.D. . . . ," says the writer just quoted, "comes into teaching looking for converts."[6]

This attitude is in part a revulsion from the excess of loose, "viewy" courses, which came in with the adoption of the casual style and the small discussion group. But in the main the new professional outlook is an emanation from science, which is then plagued by the question: "Which do we want to do—'make converts' or educate the laity?" The conditions are not interchangeable; the dilemma is a fact, not a matter of opinion. Suppose a college student not an intending scientist, who is unabashed— perhaps even cheerful—at the idea of the required two years of science. The first year he chooses an introductory course in a physical or biological science for which his high school preparation is useless. Designed for professionals, the college course is the first of a sequence of six which slowly lays the foundations of technique for research and discovery. The college student is falsely put in the position of a first-year law student who, knowing why mastery is expected of the practitioner that he wants to be, buckles down to the task: he has a motive.

The student laboring under the science requirement has no such motive, though he may have an interest in "science" or even a science. Now, no one dreams of proposing that a panorama of all science or one science be given in one year. But no requirement which is a proposal for teaching science to a scientific culture can afford to ignore the difference of pace and treatment between training an apprentice and informing a serious student, the one during six years or more, the other for one or two.

This is the difference President Conant grasped when after the second world war he designed a course not for scientists but for college students, and wrote the book *On Understanding Science* to make his point. The point is that understanding science is not the same thing as knowing scientific techniques and scientific facts. Understanding means having a "feel for the Tactics and Strategy of Science."[7] The book illustrates by famous instances what is meant by these analogies, and it thereby shows what a genuinely *introductory* course might be. Future scientists themselves could take it with profit, as a means of knowing consciously and critically what they are in fact doing when they are "doing science."

In college at least, the science that is done according to the professional formula follows the standards of the profession. I shall return to what this means for our imaginary college student if during the second year of the requirement he takes mathematics. At this point I must explain why the preparation he received in the lower schools was of no use for his first elective. Much of what is called science in the public schools is something else, semi-vocational, semi-domesticating in purpose. The workings of the gas engine and the vacuum tube are demonstrated to boys and girls, with no real hope that the knowledge will be formative as well as informative. In any case the confusion between science and techne continues. A "serious science course," in the schoolteacher's mind, means elementary physics or chemistry. It is taught, as far as substance goes, very much as it was seventy-five years ago.

The main addition is verbiage about the "socio-economic implications of science." The teacher's syllabus points out his duty to make science contribute to the pupil's "growth in intellectual honesty and moral and spiritual values."* State boards and other professional groups frequently issue such lists of "objectives" and "criteria for evaluation," but by the usual educationist perversion these are parodies of human thought. The able teachers here and there are overborne by this flood of false philosophy and piety which is twice a calamity in that it denatures teaching and casts discredit on broad views, however intelligent and sober.

What emerges from the confusion is the great doubt: can science

* From the informative and neutral survey of the subject made by Nathan S. Washton, of Queens College: *Science Teaching in the Secondary School*, New York, 1961. Compare with the jargon and moralism quoted at length by Dr. Washton the simplicity and good sense of the Leverhulme Report, *The Complete Scientist*, London, Oxford Press, 1961.

be taught except by teaching its principles? And do these principles exist in teachable form? The fear is that principles without facts will prove empty, like the traditional lab work, which so often leaves no trace of information or discipline. Knowledge of facts must go with discipline through principles; no choice between them will do. A great biologist of the last generation, J. S. Haldane, believed that the increasing neglect of this maxim harmed medical training: "On their practical therapeutic side medicine and surgery are . . . left without definite general principles, or 'institutes,' to use the old-fashioned term. Physics, chemistry, normal and pathological anatomy, bacteriology, and physiology are all of the greatest help in diagnosis and in the avoidance of disease and injury, but among them they do not supply the missing 'institutes'; and the consequences of this defect are very serious."[8]

Framing such "institutes" cannot be done once for all in any science; and it does not get done automatically. First-class minds in each science who are also familiar with the realities of teaching are needed; their task being to make their subject coherent and clear for the use of other teachers, wasting no time in defining objectives of "growth" and "creativity" and other virtues that are within no one's control.*

Failing this, the prospect is poor of teaching anyone—teachers as well as students—what any science is about. The contents even of subspecialties are growing relentlessly and with them the period of minimum training. The Ph.D. is no longer the end of the road but a stage. Post-doctoral fellows swarm in every laboratory, expensive hangers-on at every great center of learning. In medicine, eight years from the student's admission to the course is a common span before practice. The latest cliché is that "it is in those last years the subject really comes alive." Nevertheless, adding facts and subdividing sciences can mean diminishing returns to the man and the

---

* Among the fundamental sciences, chemistry has made the most progress: for the secondary schools, in reorganizing the subject around the principles of chemical bonding and molecular geometry; for colleges and graduate schools, in specifying four main fields of study as central. See Lawrence E. Strong, "Facts, Students, Ideas," *Journal of Chemical Education*, March 1962, 126; and E. L. Haenisch and E. O. Wiig, "The American Chemical Society and Training in Chemistry," *Association of American Colleges Bulletin*, May 1956.

nation. Ten, twelve, fifteen years of learning are as many years of forgetting too.

But an organized, manageable subject matter is not enough. The capable scientists who are to teach it with the hope of at least reproducing their professional kind must come to terms with the ugly word "pedagogy." Because the "methods" taught by teachers' colleges are rightly held in disrepute as fraudulent, it has come to be believed that pedagogy has no place, that anyone who tries will naturally master the art of teaching. There is no more mischievous lie. Good teaching requires advice, self-criticism, and guided practice.* This is especially true in the teaching of science, for reasons of temperament: the research man concentrates on the observation of things rather than of men. Yet no one can teach another without an intuitive grasp of that other's difficulty—not what is difficult in the abstract, but what puzzles *his* mind right now, about *this* trifle which will block everything else. Science teaching is often admirable in exposition but inept for the confused minds in the class.

The defect is at its worst in mathematics, where so often the very power to conceive and juggle with abstractions blots out the awareness of other minds and atrophies the faculty of speech. These lacks are especially deplorable at present, when the science has tried to reorganize itself in a "principled" way. For centuries a cornerstone of liberal education, mathematics was reduced by the advent of science into a "tool of research." It was taught by the memorizing of formulas and the solving of repetitious problems. The young were not to know what they were doing or why: algebra was simply a puberty rite. The reform of the past ten years has tried to bring mathematics back to its ancient state as an intellectual instead of a mechanical activity.

But it forgot pedagogy; that is, it overlooked the gap that always remains between the subject as it exists in the minds of its creators and the form it must assume in order to be taught. Any book, any lecture, any drawing is a distortion, and the necessary pedagogic

---

* To give a single example of pedagogic failure, most novices in the classroom or laboratory do not know anything about perception. They suppose that whatever is shown will be seen, ignoring the mind's need to be prepared for seeing. This is bad enough when the subject taught is, say, fine arts; it is still worse in the sciences, where observation calls for an uncommitted preparedness.

distortion is that which connects the ideal form of the subject with the actual capacity of the learner.

The mathematical reformers chose to overlook this difference in two ways: first, under the shadow of Whitehead and Russell's *Principia Mathematica,* they replaced "working" mathematics (or Analysis) by abstract mathematics (or the logic of mathematical forms). In so doing they discriminated against scientists and engineers who need analysis and do not intend to become mathematicians. Second, the planners made little or no provision for the transition from the old discipline to the new. The result was foreseeable: even students who had won scholarships with high test scores in mathematics failed the new courses. In some classes, from thirty to fifty per cent failed. One did not need to be a believer in statistical distribution to detect something wrong with the arrangements. Having as part of my academic duties inquired about this situation at half a dozen ranking mathematics departments, I found the trouble to be general. All but one of those in charge appeared to take the "mortality" as a fact of nature to which one had best get accustomed.

In its purest form, then, mathematics is unteachable, like art; that is, the teacher can only guide and perfect a native talent. This excludes mere intelligence, yet some instructors of mathematics none the less grant the need of at least one advanced course for non-mathematicians. But they quite spoil the concession by referring to it among themselves as "mathematics for idiots." Setting aside the scientific inaccuracy of the remark, the thought is a double-edged peril, like all condescensions. In a society that professes to be democratic every neglect of potential ability is a flaw. Nor are subjects divisible: we do not build a separate kind of architecture for non-architects or serve special roast chickens for non-cooks.

True, gradations of interest, purpose, and talent exist and must be provided for. But a scientific culture cannot rely for its leaders on the happy accident of nature. At present the schools cannot find enough competent men and women to be teachers. To discourage by contempt those who have ability but not the ability to be discoverers is to defeat one's own purpose. For the neglected "idiot" in mathematics or other sciences will drift into the lower schools where his preventable faults will be transmitted to his pupils. These in turn will be among the future graduate students, over whom the ritual groaning goes on: "Why don't the schools prepare them properly?"

To their credit, many science departments recognize this simple sequence by inducing their most experienced senior men to teach the introductory courses. But sound pedagogy should not stop with the teacher: it should inform the book. Scientific textbooks are usually painstaking and comprehensive, and if successful any factual mistakes are corrected in new editions. What is overlooked is the tactics and strategy of exposition. The friendly colleague who points out a misplaced decimal will not notice a baffling confusion of relative pronouns: he knows what the writer intends. In my experience with puzzled science students (they are often afraid to consult their science instructor) I have found that bad prose accounts for a good deal of the difficulty that is believed inherent in science.

Definitions, for example, can be so clear and impressive that they will never need to be repeated—or they can leave a perpetual doubt where no difficulty or subtlety exists. Students who have a special feeling for words will often be handicapped by reading the text carefully, for it will positively mislead them.† The ready reply that those who lack an instinct for the subject regardless of obstacles will probably not go far in it is good enough if the purpose is to select the great talents, but it misses the point of giving the educated man some sympathy with science. The national problem is not, How to teach those who need no teaching; it is: How to hold and fill the minds of those who can grasp a subject only by learning it.

When one turns the pages of the books addressed to the youngest pupils this indifference to expression appears in all its criminal folly. Nothing could more surely breed superstition in the future laity than the cocky nonsense found in most elementary books about

† An educated chemist, who worked with a committee on the framing of questions for one of the national "objective" tests, told me that his colleagues often showed impatience at his efforts to make the questions absolutely clear: "We're not testing their English, but their Chemistry!"

science. What is a thoughtful child to make of the statement that "the experimental test is the most convincing. But when experimental evidence is not available, the scientist will accept the most reasonable explanation"?[9] Again, how vicious is the false excitement aroused by brash comparisons between the lives of Richard Roe in 1760 and John Doe in 1960, the one scientifically and technologically destitute, the latter a veritable king. And how difficult to introduce sober ideas of evolution subsequently, if it is first presented in purple prose, followed by a request that the child state his acquired and inherited characteristics (*sic*). As for repeatedly linking the names of Darwin and George Stephenson, Edison and Pasteur as great scientists, all this does is prepare the citizen not for understanding science but for accepting the thought-clichés of the culture.

Some reformers, to be sure, would give up standard textbooks and anything else suggestive of study. They promote devices which make plain their authors' belief that science is extremely difficult and unpleasant, something to be got around or made palatable by dodges and sugar coatings. I do not speak of the absurd recordings to impress the memory with the binomial theorem by playing them during sleep or (taken like a laxative) just before retiring. I mean the efforts to present the fundamental sciences in lavishly, indeed distractingly, illustrated books;* or through the use of some familiar, "practical" idea as "a focus"—for example, by treating half a dozen sciences as they relate to conservation; or again, by restraining teachers, in the name of psychology, from referring in the early grades either to science or to great scientists, "lest they serve as deterrents to children, who are apt to suspect eccentricity in any form."[10]

Still more dubious is the use of television programs and so-called teaching machines. The several series of television courses on science that I have sampled were very carefully made, and this was their defect: the instructor was acting, not thinking. Every object was in place, the diagram superbly drawn. Progress was so smooth it left

---

* One scientist at least, Professor Robert Heizer, has been heard to declare that "one idea is worth a thousand pictures," and he accordingly prefers to teach by lecturing rather than by showing slides and films.

no grip for vicarious effort by the student. He did not even have time to copy the diagram, for after the demonstrator had pointed to it swiftly his calm voice moved on to another subject. Twenty-seven minutes go by too fast for developing an idea, though it seems long enough to be watching talk.

Seeing is of course indispensable to learning, particularly in science, which is of the eye. Visual aids therefore have a place in the laboratory. And most students, not being future scientists, will learn more from good films of important experiments than from their own fumbling attempts. But sometimes they must fumble too, and have a teacher who fumbles on occasion, and thinks all the time he is in class. One learns not by a photographic copying of things shown, but by an internal drama imitative of the action witnessed. When the instructor gropes for a word, corrects himself, interjects a comment or an analogy not directly called for, he gives a spectacle of man thinking which no slick film or televised show will provide.

As for teaching machines, they are (as might have been expected) misnamed. They do not teach, but serve as storage boxes for facts, hence test memory and accuracy—like the back of the algebra book where the answers are given. The grave objection to their use is that they will impose machine rigidity on the plastic mind.* The learner is all too prone to cut-and-dry the subject in the way the machine will do to perfection. The good teacher prevents in the pupil precisely what the machine requires.

Besides, if machinery were a substitute for teachers, books would have done the replacing long ago. How many children or adults who are not professional scholars can learn a subject from a book? Few, because a book does not talk or guess what the trouble is. Nor is it tactful. Likewise, a machine that will stubbornly repeat the previous question because Johnny is wrong about the next. The child knows that he knows the earlier step but stumbles over the

* The inventor's pedagogy falls, as I think, into the error of "factuality" discussed earlier. Dr. Skinner of Harvard is reported as saying that "the machine will teach the child to add, subtract, read, write, and spell and do other factual tasks. The teacher will be . . . teaching the refinements of education, the social aspects of learning, the philosophy of it and advanced thinking." This is the irradiated bread all over again. It ignores the importance of the "philosophy" of reading and writing, the "refinements of education" in arithmetic and even spelling. And when will the machine laugh at a funny mistake?

next; repetition will not help: I have known children who kicked their tutor in the shins for failing to grasp this simple point.†

The truth is that systems—in which I include teachers, programs, and books—always undervalue the resilience of the young mind and the mysterious power it has of finding easy and natural that which greater minds in the previous generation found difficult.‡ It is consequently a mistake to classify minds too soon and to work too hard at adapting subject matter to the types we think we find. The practice is only another misuse of the machine analogy. Sir Harold Nicolson, for example, believes that his "brain cells were not disposed in the right pattern for learning algebra."[11] And parents meekly accept the school's verdict about their children's "type." This rank superstition can be refuted by merely looking around. Among adults one finds not two types, but endless combinations of talents: scientific and mathematical; mathematical alone; scientific and literary (or historical); scientific and musical; scientific and pictorial; literary and pictorial but not historical; literary and mathematical; literary, pictorial, and musical separately; and so on without end.

But circumstances intervene, subjects and their teachers attract or repel, and ultimately the impression takes root that no other development of the mind was physiologically possible. It would be truer to say that science cannot be taught—at least not at present—than to say that half or two thirds or three quarters of the educable are "not made for it." The classification risks being the cause of what it predicts.*

---

† Akin to the machine is the multiple-choice test. See Dr. Banesh Hoffmann's *The Tyranny of Testing* (New York, 1962), which gives a clear and detailed account of the harm done by the so-called objective tests. In a sampling of scientific questions Dr. Hoffmann points out that superior knowledge and reasoning power are penalized, not on purpose, but as a side-effect of convenience.

‡ "Young students who learn physics today, if they are properly introduced to the subject, think of psi-functions, probabilities, and Hamiltonians without a troublesome sense of mystery. They learn to handle them with phenomenal ease and often take them to be more natural than the older mechanical models . . . our teaching methods are not as yet completely suited . . . [being] largely based on the physics of the last century." Henry Margenau, *Open Vistas*, New Haven, 1961, 193.

* A distinguished bacteriologist at the Rockefeller Institute once told me that in the bacteriology course at the university he was forty-sixth in a class of forty-six.

Clues to the circumstances that might favor the learning of science could doubtless be found in the reminiscences of scientific men. Some name a teacher or a book which set them on the path they now follow. Dr. James Watson, in the autobiographical speech I quoted, said he began as a bird-watcher, toyed with botany, and wound up as a biochemist, without feeling that he had either broken with his past or experienced a vision. That is no doubt the ideal way. But all we offer the great majority in our schools is an exposure to "the science requirement." We have seen how unsatisfactory it is in action. But even if it were good, it would not do what was hoped from it. For despite the professional clichés, it does not show by classroom example "what it means to be a scientist." It shows only what it means to teach freshmen science.

Perhaps not much else can be done at first. The beginner must learn facts and techniques, and in his contrived experiments he comes as close to genuineness as he is able to do. But this, it seems, often fails to disclose the charms of science. They might be more apparent from a recital of the growth of science than from the pointless repetition of selected fragments. In other words, the true life of science may reside less in its techniques than in its history.

As an academic subject the history of science may take the form President Conant chose in order to show the mind at grips with scientific problems—the equal importance of imagination and apparatus. Or it may take the narrative form to show how decisive results were wrested from confusion. For example, an account of the vicissitudes of the molecular theory before Avogadro's work was fully understood will give the student a richer idea of scientific work and thought than any textbook flanked by thirty weeks of laboratory exercises. Among the important truths to be gained from an historical review is that in the period of groping several hypotheses usually hold promise and are well supported by facts. Again, it is remarkable how much idea, wording, and outlook upon other parts of science contribute to the rightness of the final theory.

With such truths in mind the student can no longer think of science as his parents and his newspaper think. The stereotype of the march of science: Problem → Method → Patient Work → Solution is broken up and science re-enters the actual tradition and saga of man thinking. This is what educates; we have been told so again and again. Before Whitehead and Sarton, William James put it, as he often did, with the finality of perfect phrasing: "You can give humanistic value to almost anything by teaching it historically. Geology, economics, mechanics are humanities when taught with reference to the successive achievements of the geniuses to which these sciences owe their being. Not taught thus, literature remains grammar, art a catalogue, history a list of dates, and natural science a sheet of formulas and weights and measures."[12]

Since the second world war, the teaching of the history of science has spread. There are chairs and departments in a score of universities here and abroad, and the demand is great for young scholars to help in the teaching. Unfortunately, the itch for the prestige of research is always strongest in new fields, and this has made the best men tackle the hardest problems—the history of ancient, medieval, and early modern science. The discovery of these beginnings is slowly changing the scientific pantheon by recognizing as important a number of early workers whose names are unfamiliar. But although college students can profit from learning that mechanics was not made into a science in a single lifetime by Galileo, but grew out of sound fourteenth-century physics, the subject is not likely to enthrall undergraduates, any more than the knowledge that the Romans had other sciences besides the law.

To attract and instruct undergraduates, topics in the history of science must relate to a familiar setting which provides an opportunity to discuss the social antecedents and repercussions of science.* Without these, science is a string of experiments and printed papers stretching over the void; the men of science are once more fenced off in a corner of the world.

Nor does the historical presentation mean anecdotes from the lives of great scientists. It is not as a biographical fact that, let us say, Newton and Faraday's religious beliefs are worth comparing, but as an element in the making of science itself. Faraday, an experimenter, regarded his belief in God as an inconsistency; he put

* These are not irrelevant, as some scientists think. Ancient China, as we learn from Joseph Needham's large-scale history of its science, was ahead of the West in clockwork and other respects, but never developed its advantages for lack of a social as well as a theoretical interest.

it down to a weakness he would not defend but could forget; whereas Newton's deductive intellect and preoccupation with biblical prophecy forced him to fit incompatibles together and declare absolute space the sensorium of the Deity.

Because considerations such as these lead to metaphysics, some reformers propose the philosophy of science as the proper introduction to the science requirement. The technicality of this philosophy —to say nothing of divided opinions about it—makes it less suitable than history as a common ground for students who specialize in science and students who do not. History shows the grand design of diversity in the oneness of time.

From a beginning, therefore, which prefaced the science requirement with the history and setting of some scientific ideas, the college graduate who did not choose science as a profession could continue "reading science" as he now reads history, literary criticism, or musical biography. For his scientific purpose many readable books exist, from Whitehead's *Introduction to Mathematics* to the works of Charles Singer, William Dampier, Eric Bell, and H. T. Pledge. Books about contemporary science are numerous and some are excellently conceived. I would mention particularly the series published by members of the Massachusetts Institute of Technology, whose editor declares that "the authors have been selected both for expertness in the fields they discuss and for ability to communicate their special knowledge and their own views in an interesting way." In this apparently trite sentence, the terms "special knowledge," "their own views," and "interesting way" define a *public* conception of science which may be developing but has not yet been sanctioned among us.

To be sure, the reading of science and its history for cultivation, even if it becomes as general as a sober advocate could wish, will not dispose of all the difficulties. It will not, for instance, resolve the contradictions that the layman feels between what I have called the Two Sciences, that is, the plain materialism of everyday science and the refined "we know nothing" of Karl Popper and other philosophers of Uncertainty. No doubt the contradictions can be ironed out by a proper wording of theory, as may be seen in the paragraph that follows; but if the bulk of mankind is to grasp and apply such distinctions when the old push-pull image of machine causality is so simple, the teaching will have to be the fiercest and most unremitting since Mohammed's.

The new philosophy tells us that in the "nineteenth century natural science conceived of man as a detached spectator of an objective universe. It held the spectator-spectacle polarity to be genuine and fundamental. During the present century, discoveries concerning the nature of the atom rendered this doctrine untenable. The nucleus of a new philosophy of nature emerged with Heisenberg's principle of uncertainty, whose basic meaning implies a partial fusion of the Knower and the Known. This theory led to a mathematical formalism which, in order to attain its purpose, namely lawful description of experience, has to speak of probabilities rather than unique events. Individual events are no longer related in causal fashion . . . although in the domain of probabilities causality still reigns. Thus has been introduced another, more significant principle of division than the old spectator-spectacle bifurcation: the distinction between physical and historical reality has appeared."[13]

This historical reality, or stuff of life, thrusts other difficulties upon a scientific culture. For when that stuff is studied methodically in the scientific form of biology, psychology, and sociology, the contradictions of the two sciences reappear. Physics ought not perhaps to be the model science, but the truth is that it is taken as such by most workers in the other three disciplines I have named— and taken most often in a "nineteenth-century" way, which is mechanistic. Some students of life, it is true, circumvent mechanism, as we saw, by invoking pattern. But its role and its meaning are vague. When in addition psychoanalysis in its many forms addresses the public, its effect is to press as rigidly as Old Physics on the modern mind. The various "forces" and entities—id, ego, collective unconscious, neuroses, repressions—all these things have become as goadingly real as a toothache, as hard to the touch as a pebble in the shoe.

The making of things and forces out of ideas does not seem escapable under our present ways of teaching, by which I mean all that is written and spoken about life in its own domain. The conclusion the public is left with is that if the good people can learn science when science has learned all about the things and forces at work in "historical reality," then we shall know just what to do about our troubles. At the moment two areas of discord monopolize attention: crime and race conflict. In the scientific nations it is inevitable that "scientific ideas" should be sought or applied as remedies. I shall have more to say about the proposed marriage of science and the criminal law. Here I point to race-theorizing as an illustration of our failure to teach science. Not that science has anything to say that would settle the conflict; but that each side enlists "science" in the person of certain scientists to affirm the equality or the inferiority of this or that race.*

Both sides assume a determinism, even when professional caution makes them repudiate it. The august words "evolution," "heredity," "genes," "zygotes," and the rest command assent by their mere use, and one observes that whereas the statistics and other facts in the writings of the politically liberal will not remove one hair of prejudice's head, the contrary statistics and biological truths certified by professors of psychology and genetics give the layman pause and lend support to segregation in the decision of a federal court.†

The men of the Enlightenment who two hundred years ago hoped that the play of science upon nature would bring moral as well as physical help to man were aware that these blessings would have to come through man's better-informed intelligence; the benefits would not sprout independently like a plant from a seed. Hence our inherited faith in the teaching apparatus whose futility we must now acknowledge—at least relatively to the first hope. It is a blind faith, because everyone has been taught to praise the excellence of scientific method without ever discussing the range and character of its authority. This omission makes individual choices hard in a society as sensitive to moral imperatives

* On the liberal side, see the symposium *Race and Science* published by Unesco in 1961, and Theodosius Dobzhansky, *The Biological Basis of Human Freedom*, New York, 1956; for the opposition, read "the scientific facts of race" in Carleton Putnam, *Race and Reason*, Washington, D.C., 1961.

† See the judgment on Civil Action No. 1316 in the U.S. District Court for the Southern District of Georgia, Savannah Division, June 28, 1963.

as ours: we do not know enough to assess the knowledge which was once thought to give automatic liberation from doubt. We now see that knowledge in abundance multiplies doubts and keeps issuing threats: you must feed your baby regularly (no, only on demand); you must put fluorides in the water (it's poison, you mustn't); you must forestall tuberculosis and cancer and hence be X-rayed (no, no: you will get an overdose of radiation).

On the whole, the positive commands win over the cautious negatives. We submit. And these exhortations about health are but part of the load. Our worst fears are the vague ones that spring up like weeds between cobblestones—I mean the anxieties that flourish between our ineradicable ignorance of science and all that science none the less tells us about ourselves, our lives, our fellowmen: in a word, the vast confusion that has lately taken the name Behavioral Sciences. Though clearly emanations of science, their flourishing and their form are due to the large numbers that democracy and techne engender. And these masses remaining, as we saw, untaught, unteachable in the natural sciences, it is through the behavioral sciences that modern man grows scientific *sans le savoir*—as we must now see.

VIII

Behavioral Science

THAT part of the scientific effort which deals with the subject of Man differs from the rest in that its researchers and publicists tend to address the public directly. This is natural since what they say is easily understood, or has that appearance, and since it is meant to influence as well as instruct. General literature and the press are full of these men's findings, conversation often pivots on their has-showns, and the educated man is well acquainted with their drift without particularly studying their works.* It is in this manner that the assumptions of physical science, transferred to the study of man, have made their effortless entry into the common mind, where they no longer seem ideas on trial but "the way things are."

A mixed group of studies ranging from psychology and economics to anthropology, sociology, and (most lately) linguistics— or as Robert Merton puts it: "from the brain cortex to entire societies"—these investigations of human affairs can claim an ancestry going back as far as Thales, the geometrician who wanted to found a doctrine of human conduct on scientific fact. Aristotle, who is also in the line of descent, comes after an interval as great as that which separates the modern investigators from Rousseau, one of their unacknowledged forebears. Since Rousseau, the men and the sciences have multiplied with each generation. Under changing names, economics and sociology were the first to consider themselves independent sciences; the others followed throughout the past century: ethnology, phrenology, anthropology, philology, folklore, psychology, social statics, psychiatry.

Today a rapprochement among them is occurring under the

* E.g., the late Dr. Kinsey's reports on American sexual habits.

name of Behavioral Sciences. The label is so new (all the behavioral scientists that ever were are now living) that it does not stand as yet for any firm unity. Yet the term "behavioral" caught on quickly, in part because of the older name Behaviorism, coined by the late John B. Watson to denote his system of purely material (physiological) psychology.† Although few of the newer behavioral scientists are followers of Watson, the association of ideas is not wholly misleading: they wish to found a natural science on the measurement of human facts; he thought he had found such facts —those of mind—in man's physical nature. The material oneness of all that professes to be science is important to the behavioral branch and recently that unity has been recognized by the National Science Foundation in its awards.

Despite the newness of the name and the sanction, the assumptions of behavioral science and many of its devices go back, like so much that we think new, to the eighties and nineties of the last century. The vast army of investigators at work during the earlier part of that century had produced an advanced physical psychology, a physical anthropology, a mathematical economics, and a statistical sociology. These large subjects were in need of compression and redefinition. Emile Durkheim, beginning in 1893, set himself the task of making sociology a true science by asking and answering the question, "What Is a Social Fact?" In every society, he affirmed, there is "a certain group of phenomena which may be differentiated from those studied by the other natural sciences. When I fulfill my obligations as brother, husband, or citizen, when I execute my contracts, I perform duties which are defined, externally to myself and my acts, in law and custom. Even if they conform to my own sentiments and I feel their reality subjectively, such a reality is still objective, for I did not create them."[1] Like the physical scientist of his day, Durkheim assumed that the observer is as nothing, but that there is an object for him to study. This is the first scientific feature of the new discipline. Society, man's mind, primitive culture, language—each is the Loch Ness

† The popularity of the new term has been aided by the publicity given to the Ford Center for the Study of the Behavioral Sciences through the works of its many distinguished fellows.

monster in the lake outside; the spectators on shore point their little cameras at the ripples.

The second feature that is scientific in behavioral science is the will to measure. The pioneer of social Quantification was contemporary with Newton. The famous but little-read *Political Arithmetick* of William Petty in the late seventeenth century had few echoes. Then in the 1800s came the Belgian Quételet, the Frenchmen Poisson and Le Play, and the Scot Patrick Geddes. As early as 1830 Quételet undertook statistical studies of man. His Social Physics, covering at first such subjects as the distribution of heights and weights among the population, was to evolve into Moral Statistics, which should tell us, for example, about the propensity to crime. His contemporary, Siméon Poisson, was a probabilist who in 1837 studied the likelihood of acquittal in trials for capital crime, with a view to predicting the result.

From Durkheim's definition of social fact to our notion of role-playing and from Quételet and Poisson's work to our efforts at predicting and preventing crime, there is only the difference between individual and mass research. By 1840 statistical regularity in social actions such as suicide was established. It confirmed the positivism of Auguste Comte—the man who had invented the word "sociology"—and it reinforced the scientific validity of another social science, Economics. There must be a series of such sciences and in them lay the solution to man's troubles in society.

The state took a good while to convince;* but industry was moving—had to move—in the same direction as theoretical science: the large numbers employed in factories and mines; the need to regulate work by legislation; the complex management of railroads, hospitals, and other large urban services—all came to require counting for planning on the basis of count.

In this joint action of techne and science on society we behold one of the strongest forces remaking our present existence: from the practical purpose of finding out what large groups need or want, the road is short to the dream of guiding each individual

---

* E.g., the refusal of the Royal College of Surgeons to comply with Florence Nightingale's request for statistics of death by disease during the Crimean War.

for his own good, on the basis of fact, with the authority of science. It is the old idea of the Enlightenment, but short-circuited to act directly, thanks to machinery human and metallic. We have added greatly to our stock of words and ideas since Holbach and the Idéologues, but the underlying principles have not changed: man's acts in society are natural objects; these objects can be counted or their paths measured; and man's behavior can be improved by discovering its laws. It is this last, philanthropic purpose of behavioral science which marks it off from the pure curiosity-sating of natural science, but which also subjects it to public criticism proportional to public interest.*

At first sight the assumptions of behavioral science appear fully justified.† Since regularities are observed in the actions of men, these regularities may be deemed akin to those observed in the reactions of matter. If studied by comparable methods, they should yield comparable results. The neutral word "behavior" points to this likeness—this analogy, rather—which embraces all that "things do." What things do is a phenomenon and phenomena sooner or later can be measured. If the scientist will only limit himself to what he can see, the measurement he obtains will capture the unseen mental and social reality. It was by such an argument that the late E. L. Thorndike justified his intelligence tests: "Whatever exists at all exists in some amount. To know it thoroughly involves knowing its quantity as well as its quality. . . . A change is a difference between two conditions; each of these conditions is known to us only by the products produced by it— things made, words spoken, acts performed, and the like. To

* "What can we do to make ourselves more practical and useful fellows?" Theme of the fifty-seventh annual convention of the American Sociological Association, as summarized in the *New York Times*, Sept. 2, 1962.

† Sociology and psychology, it is well known, are often regarded with distaste by men of letters, philosophers, and some physical scientists. An important objection is the number of rival claimants in the field and also the inflated language the theorists employ. I take up these points in *The House of Intellect* and do not want to repeat here the evidence for the belief that so much "scholarship" in unintelligible shape shows a failure of intellect typical of our time. It is obvious that, more than science, social science is plagued with dubious hangers-on. In the present chapter, therefore, I deal exclusively with other aspects of the language and theory questions.

measure any of these products means to define its amount in some way so that competent persons will know how large it is, better than they would without measurement."[2]

This procedure, which is that of all behavioral science, consists of two steps—a breaking down of the situation, condition, phenomenon into parts, which are defined; and the counting or measuring of the visible clues from which the size or amount of the part can be inferred. To put it differently: as in physical science, so in behavioral science, the framing of a scientific object is done by analysis and abstraction. Analyzing intelligence, for example, may yield half a dozen components. Each is drawn away (abstracted) from the already abstract idea of intelligence and is then measured by some simple act—making marks on paper, assembling puzzles, responding more or less quickly to a signal, and so on. A score is kept for each performance relating to the abstract "power" to respond, to visualize, to compute, to construct, and the total is then said to measure the larger abstraction called intelligence.

It is assumed that studying a small abstracted morsel of the whole will not distort the results. The assumption has to be made in behavioral science, because analysis or cutting up may not be physically carried out on living men and because in the most important situations there is nothing physical to cut up. This is one reason for the multitude of conflicting names of units and elements in the science: each worker cuts up differently. And again, just as analysis is forced to be doubly abstract in the manner shown, so experiment is often doubly indirect because of man's propensity to react "unnaturally" when he knows he is undergoing experiment. It is easier to use substitutes, such as rats and chimpanzees, and most recently cockroaches, which are said to show an amazing lust for learning.

Leaving for the moment the worth of all roundabout ways, one can see how desirable it is to study man's functions as if they were parts of the material dealt with by the physical sciences. If man is mainly composed of salts and water, it would be churlish of him to deny so large—and to the owner so uninteresting—a part of himself to the biochemist. The study of man's bodily processes *as if* they occurred outside has yielded a vast amount of useful knowledge. The workings of the mind through physical responses have also been charted with great ingenuity, and we ignore the findings at our peril.

Likewise, the new ways of inquiring into tastes, habits, and opinions are valuable in a society increasingly dependent upon common action. Business and government would blunder worse than they do without frequent and accurate reports on these gross realities and temporary states of mind. The larger and more mixed the group, the more necessary it is to discover what the group in fact does. Personal impressions are misleading: the manager of the cafeteria must know how many customers prefer milk to coffee at breakfast. And so we count, average, make corrections for seasonal or other changes, and lighted by number govern our material life under techne. These numbers match—and indeed carry out on the material plane—the fundamental principle of industrial democracy: the vote is the original form of public opinion research.

As soon, however, as one goes beyond these material and superficial aspects of human action, questions arise which the student of the students of behavior must pause over. The most obvious of these questions is about the place of man within the science: is he the primary source, given as a working whole by nature, or is he an accidental compound of more simple natures? In an otherwise unimportant book, it is said casually that certain people "*showed* a higher happiness index."*[3] This way of speaking of persons betrays an important assumption of behavioral science which is typical, though often hard to isolate. First, the newly devised index, good or bad, is assumed to be an accepted convention, like the Fahrenheit scale; next, people are thought of as delicately made machines for conducting the experiment of "behavior"; finally, they are read off like a manometer.

Clearly, the stumbling-block of a science of man lies at the point where one must distinguish between what exists and what is said to be measured. In physical science the nature or reality of the object is of little importance. But when man is the object,

---

* Newspapers are full of these upside-down statements bred by "science." *The New Yorker* picked up this: "Children who fit these symptoms often yield a case history in which, etc." Nov. 13, 1960.

shrugging off his nature or reality is a form of conduct that ranges from cavalier to dangerous. And if it is argued that the indirect method of measuring Y to determine X will work with behavior as with physical events, we must be persuaded that Y is indeed linked with X.

That linkage is presupposed because human action or behavior is charged with feelings that find expression, more or less accurate, in thoughts, words, and gestures. Private motives and social bonds are made of this peculiar stuff. But what sort of thing is it? Does the love or hatred of one person for another exist like a pound of lead? If we say it exists "very much like an object," it is none the less hard to see how it can be measured as to size or amount. The visible acts of kindness or violence of the lover and hater cannot be simply counted. They are incommensurable among themselves and impossible to correlate with the varying intensities of the emotion. Would Professor Thorndike have been willing to measure the amount of his affection for each of the members of his family? We are thrown back on asking the living subject questions to which the answers are indefinite. As the victim scans the questionnaire, the promised yardstick becomes an elusive variable: "You love her—passionately—a good deal—very little—not at all?" As the subject pulls the petals off the daisy, the expert keeps score on his love index.

If, on the contrary, one limits the behavioral researcher to the canons of physical science by ruling out question and answer as inadequate, then the observer is caught in the contradiction of a science of man which denies man's chief characteristic: articulate consciousness. In short, removing man from the science of man is much harder than removing him from the science of nature. And when the legerdemain is successful, it leaves a puzzle as to what is going on, since neither practitioner nor patient is on the stage.

Equally unsatisfactory is the method of keeping man in view but splitting him into as many separate "men" as he has functions, and then dealing with each slice as if it belonged to a class of similar pieces. How does one discern them in the first place? How does one name them, classify them? The analogy with the scientific units will not hold, for one piece of lead in the laboratory rejoins and is uniform with all other pieces of lead; one molecule, atom, or particle belongs without shilly-shally to its category, from which it departs and to which it returns without consciousness or protest. But in the sciences of man the consciousness that is left behind frequently objects. Though behavioral science is new, we have a good stretch of historical experience by which to test this old game of fragmentation. After "economic man" had been split off, boxed, labeled, and told how he would infallibly behave, the subject had no difficulty disregarding the "laws" he was supposed to exhibit.

Even when undivided, man will complain of the attempt upon him, of the damage resulting from the mere belief in economic man. He calls the idea "abstract" and "unreal," and in the form of "later scholarship" he pours scorn on the earlier abstracters.* That man's manifestations of himself do not exist as separable parts is shown by the extreme difficulty of defining his powers, motives, virtues, and intellectual creations. None of the interesting things about man—love, poetry, humor, wit, despair, happiness— can be transfixed. Man's wholeness pulls him out from under any but the most figurative and temporary analysis—as when the biographer lists the traits or talents of his subject. It is not from the list that we recognize the man, nor would a number denoting size or amount opposite each item help identification.

But we shall be told that the point is not to identify or understand an individual. Rather, it is to identify and measure the perfectly clear and long-recognized human traits that are common to all mankind, forgetting single deviations and unique mixtures. The atomic analogy is then brought in to suggest that a measure of probability about large groups suffices to justify a science of behavior, the aim as always in science being to predict on the basis of the drift and tendency of things. No doubt such an argument has weight so long as the investigation is about unambiguous actions, but this only brings us back to the material world, where so many people out of every thousand regularly die or misaddress their letters. In the realm of thought, motive, feeling, and chame-

* See Mr. Peter Drucker's brilliant summary, *The End of Economic Man*, New York, 1939.

leon deeds, the mere tendency, the majority count, brings no knowledge.

For it is not only the slipperiness of the thing to be measured that nullifies the numbers, it is also the variation in the idea attached to the name it bears. After a life of uncommon difficulty and disappointment, William Hazlitt on his deathbed said, "Well, I've had a happy life!" None of Hazlitt's biographers has been able to account in full for this individual happiness rating. It obviously included experiences of joy in art and nature, sensations of power as a literary genius, that transcended the injustices and baffled passions felt by the mere man.

Now, if such a judgment is set aside as eccentric, and happiness is defined as that which embowers the great mass under the dome of the bell-shaped curve, then according to behavioral principles "reality" is reduced to the responses of the average. The science, in other words, tailors the universe to its own capacities, rather than faces the full mystery. And even in the average conception of happiness, subdivided into "parts" for the making of the questionnaire, how can we be sure that similar opinions do not conceal dissimilar ideas? And if dissimilar, what warrant is there for adding them together? Yet we are assured by a great sociologist that there would be "no difficulty about collecting happiness ratings."[4] These, if they were regularly taken like the census, would supply "an authoritative measure of our times."* This is the way in which the presence of numbers transforms the shaky estimate of a fugitive vision such as happiness into a measure, a standard, a basis of comparison—and presumably of action.

No one can say whether any good or harmful use would be made of such an index, except in after-dinner speeches, but this contemporary fancy has counterparts that directly influence lives. Consider the consequences when a university tells the nation that it is "experimenting with a predictive index combining objective academic data with certain subjective factors which can be weighted

---

* On the surface the idea is not new. In his valuable *History of European Morals* (1869), the Irish historian Lecky regretted the lack of a device to measure and compare the happiness of different eras. What he had in mind was an ideal standard, not a numerical survey. He called his invention, half ironically, a *eudaemometer*, from the Greek roots meaning "a measure of prosperity or happiness."

and assigned numbers, as an aid in forecasting academic success of applicants for admission to its undergraduate schools." What this means is that students will be admitted or rejected on the strength of figures, themselves made up of diverse guesses at merit ranging from the score on a so-called objective test to the opinion of an interviewer.

This last is conceded to be subjective, despite the figure assigned to it. But the test score is not conceded to be a measurement equally strange and more pretentious, although in testing the test it is the answers of the good students and the poor that determine the "reliable" questions. The norm within the group decides all things. At this rate, all the errors and superstitions of history could be justified, for the best minds believed in them. The reasoning is, in addition, circular, since the objective test relies in fact on someone's subjective idea of which are the good students who are entitled to disqualify a question. Similarly, in the many classes throughout the nation where grading is done according to the distribution of right answers, the good work of one's neighbor (or his disabling headache) helps to determine one's ranking from day to day. Identical grades on successive days therefore give a different measure of accomplishment, which is an odd way of reporting performance in a standard subject.

The moral is plain: our scientific culture often expects of behavioral science answers to the question What Ought to Be, and the answer often given is a composite report that conceals under figures subjective impressions and shifting measurements of What Is. Leaning on the normal distribution curve we try to compute the greatest good of the greatest number, a response to perplexity which is now a habit, and one strengthened by every extension of the behavioral search for causes.*

---

* "The study of the causes of any variable phenomenon, from the yield of wheat to the intellect of man, should be begun by the examination and measurement of the variation that presents itself . . . which leads to the concept of frequency distribution." R. A. Fisher, quoted in John I. Griffin, *Statistics: Methods and Applications*, New York, 1962, 58.

But experience forces a different conclusion: "Mathematics misapplied only makes things vaguer, mistier, than they were in the beginning . . . formidable calculations only confirm platitudes. . . . The psychologist is compelled to translate his test scores [because] words are more exact." Polly Redford, "Ambition, Distraction, Application, and Derision," *Texas Quarterly*, Summer 1963, 19.

The determinism that warrants this search rests on the axiom that man is kin with the animals and functions like a machine. The problem is to find the physical facts and social forces that drive it. It is by this kind of study that the general public is alternately alarmed and reassured, while the philosophical are alienated. Everything about these has-showns—the language, the topic, the air of discovery where there is only commonplace—repels the critical who have not yet been seduced by the rhetoric of numbers. A report that I mentioned in passing when speaking of research illustrates the kind of work I mean. Done at a large and renowned institute which nestles close to a leading university in the midwest, the project brought together businessmen, academics, and civil servants to study Originality. "Organizations," says the preamble, "including business organizations, need creative talent. In April 1958, business men and social scientists met together to examine research which had been done on creativity and conformity." This is what they found: "People high on originality are usually high on intelligence as well. . . . Original people tend to prefer the complex. . . . Original people tend to have more energy and effectiveness. . . . Group pressures inhibit originality; groups discourage deviate [sic] opinions. . . . These pressures are at work in all organizations. . . . Original people do conform to group pressures, but they conform less than unoriginal people. . . . There are different kinds of conformity behavior. The kind which occurs depends upon the personality of the individual and on the situation in which he finds himself. Support for a deviate opinion, however small, reduces the amount of conformity. . . . There are steps management can take to increase originality."*[5]

This example fairly represents what goes on at the many centers of social inquiry throughout the western world, though it is probably true that in number of organizations and range of studies the United States still outstrips the rest of the world. Doubtless the American appetite for findings was developed on Sunday sermons

---

* A simultaneous study at Yale confirmed the inhibiting effect of the group. Ninety-six Yale students tried solving three different problems by the procedure of brainstorming used in business. The lone workers used as controls produced more and better ideas. *New York Times*, March 2, 1958. Astonishment reigns on Madison Avenue.

and Chautauqua lectures, which sociological reports resemble in their threats of perdition, laying bare of motives, and imputations of sin. I might instance the recent warning to the public: "Low Echelon Women Found Neurosis-Prone"; I would add: Rise in Early Marriages Disturbing to Sociologists; Employers Found Shifting Criteria; Personal Traits Cited in Suicides; Study of Obesity Links It to Income; Jury Prejudices Laid to Eleven Causes; A Mild Neurosis May Be an Asset; Child Adjustment Linked to Friends; Unwed Mothers Share Lack of Parental Love; Cancer Linked to Living Habits; and Sociopathic Ills Called Physical. Besides such topics of wide application are the narrower commercial ones: Experts Analyze Small-Car Buying; Drivers Shifting Back to Manual Transmission; Garment Industry Idleness Traced to Romance Decline.*[6]

From the top to the bottom of society expectations are high, as we may see when we hear a distinguished research physicist urging "more behavioral research as a tool for peace" or learn that the President's Science Advisory Committee is asking the country to "Spur Behavior Studies—Asserts Knowledge Lags in Non-Economic Affairs." The committee wants reports on the use of leisure time, travel and commuting habits, occupational aspirations, the preferences of youth, and the incidence of mental disturbance.

These random examples, though they show the likeness of descent from a parent idea, are not all on the same plane of pretentiousness. Some are ordinary surveys of contemporary tastes which imply little or nothing about man's nature and are not likely to lead to laws of behavior. Commuting habits and the chances of academic success are not behavior in the same sense, and the excuse for pointing this out is that overlooking the difference entails moral damage to man. This is not merely because he is made for the time being an object (as he is willing to be, say, in surgery); the more permanent harm is caused by the fact that no science can

* The forms of behavioral inquiry carry so much authority that no subject or purpose is exempt from their application, though recently some attempted researches have been met with outcries. The state of Michigan had to stop an invidious screening of real-estate purchasers by questionnaire and test-score (May 15, 1960) and the Navy changed its tactics when objection arose to the rating of officers' wives as a clue to the husbands' suitability for advancement (March 1962).

allow its terms and hypotheses to reflect such manlike thoughts; the anthropomorphic must be rooted out.*

From this commandment come the artificial "problems" and imaginary questions that fill the professional and lay literature: "What is the relationship of various value orientations to frustration tolerance?" What indeed? Though the question makes the layman feel unsure and dizzy, he stifles doubt because of the evident industry and good intentions of the investigators. Meanwhile the verbiage of non-thought finds its way into the common mind and induces there a willingness to live life at arm's length, as in the reports. The unmanning abstraction and "objectivity" come to seem more natural, more decent and adult, than the former reality of strong feelings and sharp thoughts. Lionel Trilling has given classic expression to this difference in his parodying sentence about Romeo and Juliet: they did not fall in love but "their libidinal impulses being reciprocal, they activated their individual erotic drives and integrated them within the same frame of reference."†[7]

No one, of course, defends the practice that Trilling satirizes. His laughter is supposed to purge society of a mild absurdity. But the plain fact is that the folly goes on, expensive and unabashed. Language and "method" successfully conceal almost any "vacuous hypothesis." The phrase is that of a philosopher reviewing a large *Psychology of Thought and Judgment*, which had the laudable aim of bringing together all that the science knows about the subject. The reviewer's verdict is summed up in his conclusion upon the section of the book that deals with thinking, attention, and immediate memory: "In short," says the critic, "one can attend to

---

* According to Pavlov when he was studying appetite in dogs, there was disagreement as long as he and his fellow searchers were not purely objective: "It was then forbidden—and a list of fines was posted in the laboratory—to use psychological expressions such as 'the dog guessed' or 'decided' or 'wanted'—and immediately all the phenomena we were concerned with appeared in an entirely different light." I. Pavlov, *Les réflexes conditionnels*, Paris, 1928, 269; quoted in A. Moles, *La Création scientifique*, Geneva, 1957, 86.

† In our day, parody never exceeds actuality: "The Golden Rule is another codification of considerations which should govern our choice of actions lest we end by sub-optimizing in terms of our interpersonal objectives." D. W. Miller and M. K. Starr, *Executive Decisions and Operations Research*; reproduced in *The New Yorker* for July 21, 1962.

what one can attend to."†[8] And the upshot is to create the illusion that, painful as it may be to our dignity and desire for wholeness, this vast analysis of the self and society is storing up invaluable knowledge: "Any exposition of the principles of personnel administration which ignores the relationship between the principles and what we know about the laws of human behavior is . . . folk wisdom, not science."[9]

And while business, which rules our working lives, feels no tremor of doubt, the private life is at the mercy of equally confident advisers and analysts. The common attitude is that which the publication of Dr. Kinsey's *Sexual Behavior in the Human Male* helped to make usual and proper: anything can be studied— one has only to ask scientifically; and what results is by and large the truth—the sample tells all.

No amount of intelligent criticism can lessen a conviction so well in tune with techne and science, with the cult of research and factuality.* In vain does a professional statistician point out that Dr. Kinsey's way of sampling makes the findings invalid; in vain do literary critics (experts in the science of saying what one means) point out that Dr. Kinsey's title equates men living in one part of North America today with "the human male"; in vain again does Dr. Karl Menninger, whose scientific qualifications are not in doubt, undermine Dr. Kinsey's very definition of sex by reminding us that "the orgasm of a terrified soldier in battle, that of a loving husband in the arms of his wife, that of a desperate homosexual

† The merry-go-round does not bring into view only pseudo-concepts. It works also in the research about the actions of men and animals. For example: "Chimpanzee selection of larger sizes of food pieces in a direct choice situation is mediated by visual size perception. The mechanisms of selection appear to be similar to human perception of the same stimulus objects." Emil W. Menzel, Jr., *Science*, May 20, 1960, 1527-8.

* The one time I had the pleasure of meeting the late Dr. Kinsey, at a small party, I was struck by his charm, which went with a naïve eighteenth-century faith in the automatic action of scientific fact. He was then working on the incidence and characteristics of homosexuality and he believed that "once the facts were known," the harsh legislation against the practice would be generally repealed. A few weeks later, he was called in as an expert witness in his home state of Indiana for the defense of several men and boys accused of engaging in a sexual session with two girls. Dr. Kinsey thought it a persuasive argument that the numerical ratio of the parties did not deviate very far from that of the unmarried of each sex within the state.

trying to prove his masculinity, and that of a violent and sadistic brute raping a child are not the same phenomenon. The muscles and nerves and secretions may be the same, but the orgasms are not the same, and the sexuality is not the same. They may add up to the same numbers on an adding machine, but they don't add up to significant totals in human life."[10] All these objections may be understood and even approved; they do not change the underlying faith by which we are possessed.

The earnest people who, acting upon that faith, translate folk wisdom into science by elaborate verbalism and the submitting of problems to Yale students and chimpanzees, would no doubt be surprised to hear that their minds were laboring under the influence not only of Pavlov and Watson, but also of Karl Marx. Yet this is so, and the high esteem in which Marx is still held by social scientists who repudiate his politics confirms the connection. He is revered as the philosopher of the scientific method of social analysis. This reputation is deserved. His theoretical works show his fondness for "objectivation" through verbal formulas, and most of the time he was convinced that man is wholly determined by "his society." As the analyst of that society Marx believed he was teaching it only the lessons it contained. No less important, Marx's elaborative, German-trained academic mind fastened upon the public as well as the social sciences of our day a sinister intellectualist dogma to which the socialist world bears witness, the dogma that social improvement is a relatively simple thing requiring a complicated theory behind it; whereas it is in fact a terribly complicated thing which requires a relatively simple theory behind it. Our behavioral researchers need not be Marxists to work and waste our substance at the complicated theory.

As a science, behavioral research is naturally imperialistic and recognizes no restrictions.* Limits would be illogical, since the postulate that whatever exists can be measured implies the sameness of all existences: there is no dividing line. And without it there is

---

* "There's nothing in any day's issue of the *New York Times* that doesn't belong in the field of sociology." Paul Lazarsfeld, leader of American quantitative sociology, interviewed in *New York Times*, Sept. 2, 1962.

no protection for the self, which is being assailed, colonized, dismembered from all sides. It is not enough to repeat this: the relentless drive to de-anthropomorphize—I shall hereafter say *unman*—must be shown, or as the behavioral scientists would like to have it put, the mechanism must be demonstrated.

The first stage consists in isolating the subject. Norbert Wiener, who is no skeptic about science, pointed out in discussing cybernetics that "all the great successes in precise science have been made in fields where there is a certain high degree of isolation of the phenomenon from the observer."[11] This is true of astronomy and of atomic physics. When we invade the atom, what we do influences many individual particles, and this effect "is great *from the point of view of that particle*. However, we do not live on the scale of the particle concerned, either in space or time." Atom and star are equally and safely isolated. But the social scientist lives close to, and in scale with, his subject matter, and to create distance he must pretend, like Pavlov, that dogs and men do not want, or guess, or decide. The dog can stand this as well as the atomic particle, because he is not going to read the findings. But man is suggestible and interested in learning about himself. Trusting science, he absorbs its definitions and soon begins to live up to what they show. Just as nature imitates art, so human nature imitates science.

Durkheim, as we saw, defined a social fact as an event that is independent of the individual because he did not create it; he is only imitating other citizens, brothers, criminals. Certainly his acts are real and imitative too, which is one meaning of social. But suppose these acts exist only because the living man, aided by his contemporaries, creates the facts anew at each occurrence? Without this continual re-creation they would not subsist and society would collapse, as has happened to tribes that lost faith in *their* social facts on contact with alien customs. The religious festivals of ancient Greece are no more, whereas Mount Olympus is still there, independent of anyone's faith and action. Hence the two "phenomena" are not external to man in the same sense, and there is danger in straining to consider them the same.

The straining is toward viewing all one's activities as processes.

The person vanishes and leaves only the seat of a process, or at best the observer of an action. The observer cannot also be an actor, for the action or process is all there is in the situation being observed. Hence it becomes sinful to agree with the philosopher from whom I borrow the convenient distinction when he says: "there are two modes of behavior: the one illustrated by the astronomer when he is astronomizing; the other illustrated by the solar system—and by his brain and nervous system."[12] Yet what people say and read shows that few any longer make this distinction: everybody refers to his "thinking process"; the doctor tells us that we suffer from "a disease of the psycho-biological unit"; and *The Science of Culture* by Professor Leslie White teaches that "Human beings are merely the instruments through which cultures express themselves."†

The corollary to believing that our actions are processes is the belief that effort, application, will power, moral conscience count for little in comparison with obscure, irresistible forces. This hypothesis is made plausible by our pain under the "pressures" of industrial democracy and by the evidences of poverty, ignorance, crime, which seem to prove that "society is the cause." Those who professionally help the unfortunate use no other hypothesis, by a reasoning which is generous but circular: "Perhaps society should also share the blame [in suicide] for failing to equip the individual with a sense of inner security and worth."[13] The notion that a person does not merely receive and suffer experience, but "takes" it in some way, so that comparable conditions lead to different outcomes, is incompatible with a deterministic science, in which men and conditions interact like the pieces of the solar system or the chemicals in a beaker.

That a belief in this determinism or in its opposite could be an active element is even more discredited. There might be little to object to in this flouting of common experience if people in general had become cheerful fatalists, but they are on the contrary miserable, appalled by the moral failures around them, and yet driven by the moral imperatives of a scientific culture to apply to

† P. 148. Note also how readily the public has taken to the belief in Hidden Persuaders and SM (subconscious motivation), that is, in man the puppet.

decadence the mechanical aids derived from studies of human behavior.

It is only fair to add that in exchange for the loss of self-confidence, the studies and findings give back a degree of vicarious certainty. Although to deny purpose in *man* while demonstrating the secret springs of action is inevitably to weaken the idea of purpose in *men,* that idea seems less necessary when one learns that purpose is lodged in the group. A substitute for self-reliance is to be had in conformity. The rhetoric of science—tables, graphs, decimals—sustains the belief in the objective, human reality, which is to say the nearness of a thousand subjective opinions. One man's view is anthropomorphic and lonely; the community's is as clear and companionable as the moon. In the arms of this collective power, we are content enough with our self-distrust. With the manners of a masochist we automatically begin by denying our spontaneous thought: "It's only my idea." The feeling is not that a solitary mind may be wrong, but that it must be.

It is at this point that the permanent halving of the self begins. The various sensations of being the playground of forces, of having no separate purpose, of being saved from isolation by merging with the group, are topped by the knowledge that what is happening to oneself is not a unique event but only another case of some well-charted regularity of nature. So far as this feeling prevails, there is no use worrying at the thought of political 1984s: we have already inflicted them on ourselves through the fallout of behavioral postulates. Like that of the bomb, the result is to destroy the shape of man, in the same imperceptible way.

Consider, for example, the after-effects of such a work as that published a few years ago on *Premarital Dating Behavior.*[14] We begin with the apparatus of questions and definitions, which includes a table of "stages" in the behavior under study. Stage A is: "No dates within specified period"; stage B: "No physical contact or only holding of hands"; and so on with increasing embellishments up to sexual intercourse, which is stage F. Some of the stages, we learn, are not mutually exclusive. Then come the statistical returns; the remarks quoted from the subjects, with commentary by the investigator; the mean and median frequencies; the correlations and references to findings by earlier inquirers; the comparisons of "current" and "lifetime" behavior—all this as detached as an insect study, except now and then when a sad or touching remark by one of the youths ("My, what a difference a car does make!") reminds us that we are dealing with human beings.

It may be, and one does hope, that those interviewed never read the report of their confessions. But by osmosis the thoughts of everyone in the nation are colored by such findings. And I leave it to the reader's imagination to decide whether the outcome, for any person reared in such forms of thought, is not a profound deprivation; a loss not cured, but embittered by his knowledge that this is stage F and that its attainment has been through a series of lettered symbols. One thinks of a domestic episode in the life of Tristram Shandy's father, and one passes on, head averted, to think of Lucretius' invocation to Venus Genetrix and to wonder whether three thousand years have passed merely to bring us to a sophisticated kind of groping for bits of life, that we may play with them in frivolous solemnity.

Modern society professes much concern for the dignity of the individual and the sanctity of sexual relations, but that this dignity is fragile and those relations delicate is never thought to be affected by our ideas, our high scientific ideas—as if we were a people without customs.* Suffering in our nerves and adrift in our minds, it is no wonder we are beset with sex and sexual problems. For sex is objective, love subjective. You can *see* and therefore prove that the *sexual process* is going on; whereas love does not rank as behavior.† To say this is to say again in another form that man is an animal, somewhat complicated at the top. But this plausible assumption is not stable: the logic and the results of pure science force it to become: man is a machine, as good men cheerfully be-

* The efforts to remind the world of these facts are pathetic and inadequate: "Changes Sought in Sex Attitudes—Educators Urge Stress on Dignity of Individual." And below: "In the case of a young woman who believes that the quickest way to marriage is to get pregnant, it was suggested that she might consider the heavy burdens she would place on a boy trying to complete his education." *New York Times*, Aug. 11, 1963.

† What Kinsey counted was "outlets." Love is of course "recognized" in sociology, but as a convenient name for the desire to marry, not as an activity in itself. See William J. Goode, "The Theoretical Importance of Love," which deals with its integration into the class structure. *American Sociological Review*, February 1959, 38 ff.

lieved a hundred years ago.† And now that we have far more complex machines, buzzing with feedback and self-controls, it would seem the simple honest course for behavioral science to define man as a self-programing Univac.* Everybody would then know where everybody stands.

For the question, once more, is not about what may be said for or against a tenable hypothesis within a professional scheme of work; the question is what unforeseen changes in action, thought, and feeling occur in modern man as a result of the diffused behavioral outlook. Its latest device is to seize on the essence of social actions by representing them as "role-playing." What can this do but destroy institutions while rendering the self odious— "am I a bishop or only acting bishoply?" Whereas the student of behavior formerly permitted himself to break up the living subject because it was not manageable whole, now he relies on man's adaptability to the demands of life for supplying science with the ready-made analysis of "roles." The various selves are interpreted as so many stances, thus wiping out at a stroke the moral distinction between candor and hypocrisy—to say nothing of the distinction between acting and the other professions.

The entire passivity of the public which allows such notions to gain currency may be thought due to the obscure purpose and language in which they appear: "Another consequence of abandoning the postulate of consensus in role definition that deserves exploration lies in the implications it has for explaining different behaviors of incumbents of the same position." The attempt is to account scientifically for differences among school superintendents, whom the reasonable man would otherwise suppose identical. The 379-page book studies the "impact of consensus on role definition,"

† Always courageous, Huxley led the van by declaring: "I hold, with the Materialist, that the human body is a machine, all the operations of which will sooner or later be explained on physical principles. . . ." This he follows with: ". . . consciousness would appear to be related to the mechanism of their bodies . . . as the steam whistle which accompanies the working of a locomotive engine [and] is without influence on its machinery." *Macmillan's,* May 1870, 78; *Fortnightly,* Nov. 1, 1874, 575.

* We have already acquired the habit of classifying ideas in the syllogistic way of the digital computer: fiction and nonfiction, native and non-native, white and non-white; and likewise: non-teenager, non-affectionate, and so on.

and gives "a penetrating treatment of conformity to expectations."[15] It would be a mistake to imagine that such a stream of
words from the behavioral academy does not quickly affect the
common mind. The toleration of the advice, sought and unsought,
that pours upon us from all the psychologists and social workers
and criminologists and counselors of every stripe testifies to a
previous brainwashing in a medium of science more concentrated
than any which the physical scientists could have prepared.

True, the doctrines and catchwords come and go, but nothing is
lost; every affectation of wisdom leaves its mark, and role-playing
is destined to a fruitful career in our psyche. For it is the form of
the device as much as its contents that confirms man in his
masochism and moral humiliation. The words "process," "stage,"
"role" are enough to prove to him that nothing is what it seems
and that he is controlled by blind forces within and without. As he
reads the books tossed at him over the whitewashed walls of the
temple of science, he learns that all he thinks is wrong, that mental
health, love and marriage, child rearing, poverty and crime,
business success and art appreciation, differ from what anyone
has ever thought or can now perceive unaided. They are entities,
or as he says "concepts," and infinitely intricate: only his ignorance
is plain. As his oversolicited wits flounder among the jargons and
the figures, he is sure of but one thing, which is that his mind
and self must be brushed aside before he can understand the
world.

It is a tribute to behavorial science that it has persuaded the
western world of the old maxim: All is appearance, reality is
nought. The success of this crusade and the extent of our woes may
make our posterity think of the similarly contrived anguish of
many nineteenth-century minds at a loss for the true way, from an
excess of evangelical doctrine. Both situations reflect the spread
not of light, but of the loquacious theology in power.

I speak of the extent of our woes, for it is evident that the
public does not lift its voice in songs of joy at the benefits it
receives from its new science. On the contrary, not a day passes
without the complaint that our present low estate, morally, socially,

and psychologically, is due to the failure of the sciences of man to progress like the physical sciences. This is one of the great common-places of the age. It flourishes unchallenged because its meaning is not even thought on—a measure of the strength in the ideal of behavioral science.

To get rid of the platitude one must first dispose of the argument that the social sciences, which have recently conceived the hope of forming one grand behavioral science, are young. This assumes that it is the passage of time which explains the success of natural science. There is in fact no such thing as a young science; there may be a small science or a short science, such as mathematics in Greek times or chemistry in the early nineteenth century. What was known was little, but every bit of it was genuine in form and most of it solid in contents. The social sciences today have yet to show one universal element or controlling "law," one unit of measurement, one exactly plotted universal variable, or one invariant relation. The ologies in their latest avatar are not so much young as unborn. I am of course speaking of the social sciences as quantitative and predictive disciplines having general truth, not of the local descriptions and statistics found in sociology, economics, anthropology, and psychology, which were well advanced in these ways as far back as 1890.* If one had to characterize the present effort, with no wish to impugn the good faith, the skill, the patience of its enterprisers, one would have to call it, not a pseudo-science, since that term should be reserved for quackery—but an "after-science," on the model of the German *Aberglaube*.

Indeed, the idea, though not the word, of this criticism was advanced some years ago by a German practitioner of the craft, Dr. Ralf Dahrendorf of Hamburg. In an article entitled "Out of Utopia," he made the point that in the effort to reach science, recent

---

* Messrs. Bernard Berelson and Gary A. Steiner have recently issued under the title *Human Behavior* (New York, 1964) an "Inventory of Scientific Findings" which lists 1,042 such findings. Those that are not mere local counts—how many students in a given region plan to go to college—are so abstract and general as to say nothing new, e.g., "power is a function of social structure." The chief value of the book is in the conclusion of the less sanguine of the two authors, who opines that the science gives "too much respect for insights that are commonplace . . . too little regard for the learning of the past, far too much jargon" (pp. 12, 662). In a word, what is true is not new; what is new is not true.

sociological theorists have, like utopians, isolated their ideas from historical reality. Their abstractions, related in imaginary "models" akin to a system in physics, can bear no relation to empirical reality and in fact betray minds having "a minimal concern with riddles of experience."[16] In these gymnastics one finds the parallel to that one of the Two Sciences which is pure and meaningless quest. The second one seems equally futile to our critic: "From the point of view of concern with problems, there is very little to choose between *The Social System* and the ever-increasing number of un-doubtedly well-documented Ph.D. theses on such subjects as 'The Social Structure of a Hospital,' 'The Role of the Professional Football Player,' and 'Family Relations in a New York Suburb.' . . . Topics chosen because nobody has studied them before, or for some other random reason, are not problems."[17]

So much for the progress of the young science. The much more dangerous superstition abroad today is that if behavioral science did progress our problems could be solved. This plausible proposition is fraught with calamity. Our "problems," so-called, are due in part to the willfulness, stupidity, greed, and fears of men; in part also to the presence in life of real choices and hence of irreducible conflicts: Washington in 1774 was not willful, stupid, greedy, or afraid; he simply preferred independence and was "a problem" to the British. He did not need to be cured or saved, but satisfied —like his opponents. How could behavioral science have helped— and on which side?

In other words, men singly and in groups have notions and wills and make demands on life. These wills resist coercion just as these ideas resist complete congruence with others. Both will and mind must somehow be got round or caught napping before they will yield to the self-interest or imposed purpose of another. Accord-ingly, leaving aside simple persuasion, which needs no help, what a science of man must do to "solve problems" is to discover means to manipulate wills and minds. When the physician strikes a light blow below the kneecap with a little mallet, we cannot help our-selves: the leg leaps forward. The knee jerk is most valuable in neurological diagnosis. But suppose that by tapping another joint or emitting a magic word on a certain pitch, we could make everybody leap forward to vote the Republican ticket. The secret of these discoveries, the exact formula, would be the answer to what the yearners for behavioral prediction and control demand. It would be literally the knowledge that is power.

For that knowledge to be useful it would have to remain a secret in the hands of the discoverers and be sparingly used. For having been coerced once by surprise, the passions of the many would seek ways to insure that the formula brought diminishing returns and the manipulators paid for their success. This situation —hardly one of social peace—is no fanciful projection. Many energies today, and especially the philanthropic, are bent upon manipulation. Our equipment—techne—points the way: truth serum, the compulsory psychoanalysis of groups, the deceiving of patients and subjects in research for medicine and social psychology, the use of wiretapping to find out how a jury works or a citizen conducts his life—all are steps toward that understanding which shall give control and promises progress in the science of man.

If asked whether in the light of present practices one should fear science, the answer is: No. But fear men. The double temptation of machines and formulas is too much for most intellects to resist. Everybody thinks that with the appropriate power-and-device he can circumvent the animate or inanimate obstacles in his way. The student hopes to learn a foreign language by having gramophone records play in his sleep; the medical quack uses electronic Illness Meters;* others consult computers to predict the decisions of the Supreme Court; and all large institutions find themselves forced to treat their people as once convicts alone were treated—by number, secret reports on behavior, and cumulative evidence of past deeds.

There is no need to impute evil motives for the evil to be threatening. Any social scientist in the grip of factuality—that is, who thinks that findings dictate choices—is potentially a manipulator. The presence of many such workers among us, the likelihood of their desire to act on what they find—indeed the bare possibility

* "United States marshals raided chiropractors' offices . . . yesterday and seized five electronic machines purporting to diagnose scores of serious illnesses. The complaint contained no allegation that the . . . chiropractors knew they were useless." *New York Times,* Aug. 14, 1962.

of this action—leaves modern man much heart-searching to do. Is this knowledge he can handle? Whoever wants the behavioral sciences to catch up with the physical sciences probably thinks that "when the facts are known" they will influence reasonable men by sheer weight of evidence. But the appendix to this is that when they fail to influence there will be every "scientific" justification for seeing to it that they do.† Again, when our contemporaries regret that we are not able to predict and control the behavior of man, who is "we"? The formulas of control would be like those of the bomb, something nations would fight to possess that they might use it in offense or defense; and the situation would soon revert to one in which people would say that the sciences of man had not kept pace with the science of nature. When we see how ineffectual is our control of manufactured foods, drugs, and implements, about which our knowledge is ample, it argues not a vivid but a weak imagination to speak of control over man through science.

Finally, the man who yearns for a science to control man must ask himself a philosophical question. If the study of man takes the form of a science exclusively, it will show that all events in society happen by chance: the regularities are probabilities, unconcerned with the individual, whose will and self-control are illusions. Does this not mean that science brings back into our world the randomness and indifference which man strives to remove by purpose and standards?* Morally, statistical living is a regression.‡ Science may indeed be used, like anything else, for moral and spiritual ends, but man must be aware that the assumptions—and hence

† The position is, again, the one Marx took early in life: "We expose new principles to the world out of the principles of the world itself . . . we explain to it only the real object for which it struggles, and consciousness is a thing it must acquire even if it objects to it." Introduction to the *Deutsch-französische Jahrbücher*, Paris, 1843.

* A French writer on probability has been at pains to point out (*J. Duclaux, La Science de l'incertitude*, Paris, 1959) that there is actually no chance that two people in love should be "made for each other." Even if this proof were coupled with the proof that arranged marriages turn out as well as free choices, which is probable, what relevance has either thought to the consciousness of love, marriage, and freedom?

‡ A young woman interviewed after the brutal murder of two girls in her neighborhood said she was sure a like event "could happen anywhere in New York. I only hope I'm on the right side of the statistics." *New York Times*, Aug. 31, 1963.

the results—of science come out of moral presuppositions about the universe and the self. If we go for lessons on life to the laboratory and the cage as Wordsworth went to vernal woods, we ought not to complain if the replies differ. But while we become what we collectively think we are (resembling in this the young scientist in Hugh Davies' fantasy who, after studying rat behavior in the sewers of London, decides to leave his friends and join the fast-evolving race below)[18] we can discern among the students of man's physical nature a new spirit of self-criticism that may in time sober down behavioral research. Reversing, in fact, the behavioral assumption, neurologists maintain that "it is behavior as much as structure that determines the mechanics of the body";[19] and conclude that the "science" of the text book distorts reality by the very fact that it is written down and that it records a hypothetical "average" of medical practice.[20]

On their side, biologists have noted a greater variation in animal traits than the headlong faith of behavioral scientists has supposed. Some psychologists are guided by "a continuing realization . . . that perhaps the white rat cannot reveal everything there is to know about behavior";[21] the evidence against the idea that rat= man piles up: the vitamins and sulfonamides that affect rodents in certain ways affect human beings differently; dental research on mice proved delusive; the tubercle bacilli in birds and mammals are unlike; the rat, king of creation, scorns consistency: he will not respond to diphtheria toxin identically in different parts of the world.[22] These discrepancies show how necessary it is for science to continue devouring its own offspring. From these particular revisions of doctrine, one learns at least the great lesson of the fluidity —almost the evasiveness—of life, its susceptibility to change under influences mental and physical.

Every time that thoughtful men have pondered the results of their physical research into man they have been forced to conclude that the tendency of living beings is at the opposite from the mechanical. How can man be classed as a machine until we have a theory of disease? That is, until we are sure that the bodily failure represented by a broken leg is the prototype of every illness? Far from behaving (or should one say *behavioring*?) with the regular intelligibility of a clock, the bent of the living and of man in particular is to *mis*behave, in all senses of the word—from developing allergies, which make poison out of delicacies, to committing crimes which, as in saints and statesmen, can later seem the highest wisdom. It is even proved by research that man must have his ration of dreaming, that is, of irregular and inaccurate thinking.[23] These facts of experience require that any science of the regularities of behavior be always qualified and admonished by another discipline, a learned lore of misbehavior.

IX

Misbehavioral Science

IN A farcical parody of the guild—obviously the work of a member in good standing—one reads how the behavioral scientist Socinus Grelot paves the way for his new science of Neurosociology by reminding us that "the man in the street will say: 'I feel it in my bones . . . but the hard-headed osteopath can ask, rightly: 'Do these bones live?' And without the nerve tendrils, clearly they do not. . . . All twitching begins in the nerves, and it is to the neuroses that we must turn for knowledge."[1]

For us who have tried to see science whole, the time has come to turn our backs on the nerves (so to speak) and look for knowledge elsewhere. For the excellent reason adduced by Grelot—whose pen name hints at a Dr. Bell—it is not in our bones that we find this knowledge, but in the elements of our culture I have chosen to call Misbehavioral Science. In calling them so, I am playing false to the definition of science I gave in Chapter II and using "science," just once, in the general and still common European meaning of exact knowledge. My purpose is to make clear the contrast between the study of the regularities in life, which engrosses the student of behavior, and the study of the irregularities, which concern the student of misbehavior.

To blur the contrast and then represent science and the humanities as competing and mutually nullifying is a modern error. Each can legitimately claim a portion of the specifically human as its proper object. If one wished to speak precisely, one would compare the Humanities, meaning by the term knowledge of conscious man and his acts, and the Objectivities, or knowledge of things and processes. Unfortunately, the connotations of "objective" suggest a monopoly of reality and truth, so the term would only increase people's doubts as to whether their thoughts and feelings are real. This absurd but powerful suspicion should perhaps compel us to say: "The Regularities" and "The Irregularities," which would have the merit of being a fresh and uncontrived pair of words. Yet they will not do either. They vaguely echo "The Proprieties" and suggest genteel occupations. So it is best to continue using "The Humanities," though with the amplified meaning I have just now tried to give them. This once clear, it is also best not to depart from the earlier definition of "science" as the study of objects, tangible or abstract, through analysis and measurement, for prediction, control, and other intellectual satisfactions.

In common speech the humanities are not an exact counterpart of science, because science claims the complete cosmos, while the humanities are but a few disciplines of art and about art, which look as if they could be easily dispensed with. The realm that more exactly balances the scientific and cannot be dismissed as optional is history. I do not mean by history only the narrative of the past and the scholarship it is based on, but all the records of man, including the scientific, since science is part of man's history. Likewise art, literature, religion, language, philosophy, and the law. These activities are mirrored in written history, in archaeology, and indeed, in those parts of the ologies that are not science in form or purpose.

One cannot always draw a sharp line between works of history and of science, because history often reports scientific truths, just as scientific findings often refer to unique historical events. But the drift and purport of each kind of work is usually unmistakable in tone and quality. For as Henry Margenau puts it, the historical universe is that which hits us directly and whether we like it or not: "Items that partake of the character of immediacy, spontaneity, and coerciveness in our experience will be said to compose historical reality."[2]

Out of this reality come the objects and subjects of the arts and of historical studies. These arts and studies have this in common, that they reproduce the element of time and that most of them use words. And their common message and device of expression is about man in his irregularities, rendered through particulars. Not formulas applicable anywhere at any time, but a statement, fictional or literal, which says: "It was thus." When Caesar reports that Gaul is divided into three parts and goes on to name the tribes inhabiting each, we learn what has been and need not have been and did not remain so. Historic Gaul no longer exists, though the natural land is still there; there are no more Sequani, though men still live in their regions. The existence of Gaul, as of Caesar, is unique and unrepeatable. The reason for knowing about it is that the knowledge satisfies curiosity, as in science, and also that it leads to an understanding of men.

The readers of history and literature, for instance, do not need to gather in teams to reach the conclusions about creativeness and conformity that were laboriously rehashed at Ann Arbor.* They know that Originality is Misbehavior, and they infer instantaneously all the experts' corollaries about the way institutions are likely to treat those who misbehave in that way. This knowledge, incidentally, is just as "objective" as that produced by committees' shuffling through "studies"; and it is solid truth even though it is the thought of "only one man." His knowledge, moreover, has been acquired at large and not at a conference: he has lived in historical reality and read the fiction and biography arising out of it.

This brief reminder should suffice to disqualify the common antithesis of subjective and objective, which I have tolerated until now for expository purposes: it is badly misleading and should be dropped. All thought whatever, by a scientist or by the village idiot, is subjective in the sense that it is the activity of a living subject. Such thought turns objective the moment it becomes a considered judgment of some matter which is open to the inspection of others.† In short, Subjectivity is not idiosyncrasy; Objectivity defines a relation and not the success that attends it. A judgment can be wrong, even perverse, and none the less objective; and there are better words than "subjective" to denote error.

Again, the fact that the objects of science are conceived more

---

* See above, p.174.

† Thus a ringing in the ears caused by quinine is not objective if it is mistaken for bells; but the occurrence of such illusions is an objective fact that only a subject can discover and record.

nearly the same way by all capable minds increases the cogency or precision of the thought known as science, but it does not increase the objectivity—the link-with-object quality—of that thought.† And as the scientist creates by abstraction objects for us to think about, so does the artist or lawyer or historian produce them by concretion through various means. There is now a Fifth Symphony of Beethoven's for us to think objectively about, and a Mona Lisa. The Oregon Trail was a concourse of earth particles until by writing of its human meaning Francis Parkman made it an object of contemplation.

In choosing from the vast range of misbehavioral knowledge the few topics that follow, I do not mean to imply their pre-eminence. They are merely the ones that best show the accommodation of that knowledge to scientific culture. One might begin by questioning whether today the word "accommodation" can be used: there is confusion about the nature, role, worth, and proper contents of "the humanities" as superficially there is not about science. When the partisans of each address the others, misconceptions of purpose, resentment of power, envy of money, contempt of attitudes, blind concessions on both sides to reconcile the parties and assuage the pain, are rife. We saw before how eager some scientists are to be thought poets; but this ambition once verbally satisfied, there is no matching belief in the rigor and reliability of poetry, or indeed of the other historical disciplines. Rather, they are regarded as "inspirational" merely.

Under the weight of this usual opinion, the humanities cannot help being defensive. They concede that what they have to say is variable in truth and application; they regard the onset of behavioral science as an irresistible tide that will drown them out. Their one last-ditch justification is that to cultivate the humanities makes a man more gentle, moral, tolerant, and contented—in a

† In the historical disciplines the certainty can be equally great: "All such disciplines remain subjective in their generalizations, their syntheses are influenced by the temperament, the antecedents, the outlook of those who practice them. . . . But our stories of the past need not differ from each other in their factual contents. . . . Conflicting versions will have common elements, and on these agreement can be achieved." G. J. Renier, *History: Its Purpose and Method*, Boston, 1950, 50.

word, wise. That is how the humanities can "make a contribution toward solving our problems."

I have stated here the theme of the great "and" literature—the humanities *and* the democratic way of life; the humanities *and* the well-adjusted man; the humanities *and* creative leisure; the humanities *and* the attainment of mental health; the humanities *and* world peace. Implied is the working of cause and effect, and with it a covert appeal for support: cherish us for the aid we shall bring to democracy and peace in our time.

The trouble with this promise and argument is that they are false. If the humanities possess the powers attributed to them, why have they done so little in so long a time? They flourished in ancient Greece, they were dominant in the Renaissance, and those eras were as troubled as ours. By what magic could the humanities, which have all they can do to give shape to historical reality, also substitute for medicine and psychiatry, for juvenile courts and international organizations? The humanities draw all their substance and most of their meaning from the *in*humanity of man's life; they wallow in blood and strife and disaster. Scriptures are fierce history; the lawbooks bulge with enormities; and the poets preserve the same fiery stuff of misbehavior: The *Iliad* is not about world peace; King Lear is not the model of a well-adjusted man; *Madame Bovary* does not teach the judicious employment of leisure time. Nor is there proof that the contemplation of error and evil in history or literature has ever prevented a bad or foolish action. In short, the modern attempt to use the humanities as agents or formulas of betterment is simply a careless transfer from the scientific outlook in its behavioral form.

The ill-disguised uselessness of the historical studies—excepting perhaps the law—marks them down in a culture that measures by techne: if they are of no use, why have them? Even at their low cost they seem a waste of time, money, and energy. To this one may reply that science too is useless and even more costly.* But one

* See Chapter V and compare: "As an observable fact, the overwhelming mass of scientific thought and observation in the western world was never designed to bring, has not brought, and is highly unlikely ever to bring the slightest improvement in material standards of living." Professor John Jewkes, speech at the annual meeting of the British Association, 1959.

can do better than give a *tu quoque* argument, by showing how inept is the meaning of "useful" and "practical" as we commonly apply them to intellectual subjects. It is in fact one of the great uses of the humanities to force one to consider the idea of use. We say of poetry, for example, that being cultivated "for its own sake," it has no use, whereas engineering is pursued for machinery, which will make usable goods.

The fallacy begins with the words "for its own sake," which are ambiguous. It is not for *poetry*'s sake that it is read—that is, to do poetry some good—it is for your sake or my sake, because we like it and find it good. And that is its use. If there is an appetite for poetry, the production of poetry is as useful as the production of wedding cakes. The ranking of commodities according to need or rarity is another thing.* But that people are born in every generation to whom poetry and the arts are a necessity like food is a fact of civilized life. The demand is older than science, which now fulfills a kindred need—both equally useful, both justified as high entertainment, neither meant to become one of the bodily conveniences.

Science satisfies the mind by precision and generality; except at times of "revolution," it brings men together in an unexampled way on statements to which they agree without the need of persuasion; for as soon as they understand, they concur. As against this, literature, the law, history, philosophy, and religion divide; they arouse among their devotees argument and counterargument, distaste and disapproval. Short of war and murder, there are no hostilities comparable to those of humanists. This is because, dealing in particulars, the historical disciplines lack the geometrical qualities of science. Their terms defy sharp definition and mathematical expression—which would be an incurable defect if the humanities were competing with science in the production of transferable knowledge. But only a scientific culture tends to require that all knowledge take but one form. I say tends, for although the thoughtless nowadays see nothing wrong in the demand, our

---

* In the New York City "Redbook" for 1962 there are listed several columns of Psychologists, a good number of Public Opinion Analysts, and a score of Consulting Scientists, but only one Metaphysician and no Poets. Yet it is Robert Frost that the State Department sent to Russia for the common good.

best scientists see the difference between scientific reality and historical reality and know how science neglects the central subject of all art and historical studies: the individual.*

Through recorded or heightened or contrived examples, the historical arts give knowledge of persons. Their subject, man, being reconceived by mind and seen by imagination, but not reduced to system, has the effect of keeping always in the foreground the fact of novelty, of uniqueness, of unpredictability. The humanities dwell on what is unlike and anarchic; they find what does not conform to rule. Such is the refreshing spectacle of historical reality when ordered by art. It is the Antigone of Sophocles, who is like no other woman, which is like no other drama; the Athenian plague in Thucydides, which, though no longer namable, is vividly present and forever the past; the old woman painted by Rembrandt, whose form and features we shall never see again but in that dark glow; the adagio of Beethoven's Fourth Symphony, which arose from no formula and yields none; the Zarathustra of Nietzsche, which is an impossibility and a revelation; the lyrics of Hardy, which defy all the canons of modern diction and sentiment and prove them wrong; the languages of a thousand peoples, which are each more illogical and more subtle than the next—these are the forms and substances that the humanities present in the order of logic and veracity, combining fixed reason with wayward spirit and thus deserving the name, if only temporarily, of misbehavioral science.

About the individual, living or dead or imaginary, it is science that is inexact: its definitions are always partial (in both senses), and it fails in its own domain of prediction and control, because these apply only statistically to the mass—which is to say that science works with great accuracy by hit-and-miss. When it announces that the half-life of Radon 222 is 3.8 days, this tells us that half of any amount will have disintegrated in the time; *which* half cannot be foretold. No one cares. But if a law court jailed or hanged half the accused brought before it in any three or

* "Individual" is a science-made word denoting an object-like, statistical idea. The proper term for a human being is "self" or "person."

four days, without caring which half, there would be commotion even among scientists. The courts dare not work, much less predict, by number as science does work with and predict the emission of alpha particles. What one ought to say, therefore, is that the law is exact, but not precise; science is precise, but not exact.* In this sense all the historical disciplines from poetry to law are exact: they grip tight the single particular of the moment and never by chance mean the one next to it.

A further contrast follows. Because the objects of science are rigid and fringeless, a scientific statement of their relations is quickly assented to by all interested parties; but when a new fact collides with it, the first scheme is quickly discarded. Contrariwise, opposite opinions about historical statements persist generation after generation.† Critics praise or dislike *The Aeneid* (or *King Lear*) in our day as they did in Vergil's (or Shakespeare's) and in any century since. Nor is the debate between the good and the bad judges of poetry; conflict is inherent in the nature of particulars; they would have to be trimmed, simplified, de-concretized, to be seen alike by all observers. This is a case of our earlier distinction between the complex and the merely complicated.

Yet this being so does not warrant saying that science is perishable and art is not. For all our talk of scientific revolutions (when it is true) much remains. Euclid is still cogent and most of Newton and Mendeleev, though they are seen under larger headings which change the perspective upon details. But most of its past, science discards forever, despite its having been good science in its time. Science sheds as it grows, whereas the keepers of the historical drag a large cargo of objects considered to have transcended the limitations of their age. Thus men of the twentieth century study Latin and read Tacitus, and Continental lawyers scan Roman law, though there are no Romans left to enjoy either at first hand.

---

* Using the same words somewhat differently, Dr. Moles makes a comparison among the sciences: in his view paleontology may be called reasonably precise but not exact; acoustics is exact but not at all precise. *Op. cit.,* 116.

† "Science," says Eugene Rabinowitch, ". . . looks at its facts, concepts, and laws as a temporary and rapidly changing codex, reflecting the momentary state of affairs." "Science and Humanities in Education," A.A.U.P. Bulletin, 1958, quoted in *Bulletin of the Atomic Scientists,* November 1958, 374.

In the humanities, reason and the search for evidence are required as in science, but once again the manner of their use is irregular. Demonstration varies with each art, sometimes with each piece of work: a "convincing" line of poetry does not meet the same test as a convincing argument in politics or law, and satisfactory order in philosophy differs from its counterpart in music. Now, demonstration has been well defined as that which "generates the sentiment of rationality." In the historical disciplines, the sentiment of rationality springs from another spectacle than the scientific one of symmetry or equivalence—both halves of the equation balancing. It springs from the variety and force of the evidence that imagination brings to the reason. Imagination goes beyond what is present to the eye and calls up an indefinite number of other experiences, sensations, and ideas, with which it fills and tests the work of art, the history, the legal case. The test is that of relevance, good order, fitness for use in life—what you will: the point is that there is no single, required, definable way to respond. And so we talk of taste or judgment, we say that the appreciation of poetry is intuitive, or that the rightness of a legal decision depends on a balancing of considerations in which the mind suddenly acquiesces by reaching the same internal equilibrium.*

The clearest implication of all this is the ever-present possibility of error—in the object and in the beholder; and what is more, of unprovable error. Error of such a kind is obviously useful in a scientific culture, where one feels stifled by the multiplying safeguards against provable error. For if error in the humanities is unavoidably part of the game, then what the humanities involve is not a game; we the beholders are not players within a system of rules; we stand beyond mere deduction from the previously known. In short, error opens the door to real novelty; it is by being possibly wrong and possibly a new creation that the historical resembles (and may thus disclose) life itself.

* An English authority on the criminal law, Glanville Williams, observes that "even an experienced judge often does not arrive at his decision by way of conscious reasoning: what happens is that as he listens to the evidence, he begins to feel the answer in his bones." *The Proof of Guilt*, London, 1958, 278. These bones are obviously those rightly disallowed for scientific work by Socinus Grelot above.

If historical reality is what is immediately felt by living beings, if it is "coercive," as the philosophers say when they refer to irresistible impressions, there should be no difficulty about mankind's relation to that reality. Unfortunately, men are enormously suggestible, as well as tempted to abstraction because it is a defense against complexity: the convenient abstraction first masks the real, then is mistaken for it. In a scientific culture the power of science to manipulate through abstraction acts on the humanities, and like the moon on the sea, pulls them toward itself. The arts and their scholarship tend to mold themselves more and more on science, until the scholar or cultivator of the arts receives from them only another reminder of science. And since the humanities treat the historical realm of his actual existence, he is pushed by their alteration still farther away from the sources of life and spontaneity. I shall give some illustrative details in the next chapter. Here I merely say that modern man is without an antidote to number and abstraction, because of a lack in the department of life which belongs largely to words. This deprivation does not come from too great familiarity with science, since he lacks that too. He suffers rather from his ignorance of it, which leaves him at the mercy of its aridities.

What formerly supplied man's needs was a body of knowledge in words—the words of religion, poetry, philosophy, and the law, now displaced by self-conscious techniques and diagnoses. The trouble thus goes deeper than is imagined by those who hope for a remarriage between the humanities and science. The scientist, however absorbed in his work, remains four-fifths an ordinary man and also needs the solace of fictions other than his own. He should find them in the common air, and so should the majority, who can no more be expected to turn into professional humanists than they can hope to be working scientists. When the tradition falters and dies, teaching and preaching the humanities is futile, as we know from the schools, whether of engineering or of music, where "broadening" is required: anything "nonprofessional" is a dead letter, because it does not live as truth in the public mind.

I said that language is the chief medium of the historical and misbehavioral; it is so if only because the nonliterary arts of painting, music, and the dance are themselves subject to verbal interpretation, history, criticism, and transmission. And this reminds us that the public mind, the "common air" I referred to, lives by language too. The lack or failure in the historical realm is thus at every point a defect in language. What has happened to language, the acknowledged falling-off of the past half century, is therefore instructive about our condition.

Anyone can see the external signs and obvious causes of change: the mixing of peoples, the spate of democratic and totalitarian harangues, the burst of inventions and new sciences, the spawning of processes, abstractions, and manufactured goods, the freedom to play with language that literacy and advertising encourage—all has rendered virtually meaningless the phrase "mother tongue." The dictionary is no longer a list of words spoken by the people, but a semi-encyclopedia, in which a score of technical vocabularies lead an autonomous existence, together with initials, abbreviations, and trade names. In the holes and corners sulks the native tongue, heavily laced with slang, the jargon of war, and popular misconceptions of scientific terms. From these vocables*— nearly half a million in the latest Webster—it would be difficult to choose a central group of words and phrases whose uses and connotations form a public idiom.

This is a loss whose effects are felt as the "problem of communication" in political and business life. In private life, the tone and *interest* of conversation likewise suffer from the decreasing family likeness among the words that spring to the lips—the semi- and pseudo-technical mixed with the vaguely metaphorical. Flexibility may seem a convenience, but the user pays for it in the choppiness of his stream of thought and the crudity of the fragments of feeling it carries. For whether he knows it or not, language is the keeper and shaper of his consciousness, and he cannot habitually cast his will and fears and hopes in broken or foolish patterns of words without becoming a foolish and discontinuous mind.

It would take a treatise to discuss the unstudied ways in which

---

* "Vocables" means able to be voiced and is thus ironic here: thousands of new "words" are unspeakable and are referred to by initials, like the well-known DNA, which stands for deoxyribonucleic acid.

we record in the living languages the jerky motions of our spirit. Only a few that have relevance to scientific culture and some of its products need occupy us here.

The modern desire to be authoritative and complete in the manner of science takes the form of preferring the abstract and general to the concrete and particular. The businessman sees ahead of him a vague series of activities, meetings, paper work, and lumps them together as that which will "finalize" the deal. "Finish the business" is far too homely for his purpose. The contrast of these phrases defines a state of mind. We see it again in the classifying word, most often carelessly chosen, whose purpose is to show that the things denoted belong to the order of natural facts, capable of study. Thus one lawsuit might accidentally take a long time to settle, but when all long cases are known as "durable" they take on the air of a new product—"durable cases."

Again, a person's acts are peculiarly his own; if, therefore, we want to make them the property of the investigator they must be welded into a process or entity. Thus do desires, impulses, and temptations merge into *motivation*. Already as a child, the man showed a willingness or a reluctance to learn, which the school "handled" as "his motivation": he had it or had it not, like the measles. The critic of a novel similarly looks, not at the ideas or actions of the hero, but at his "psychology." It would be a mistake to see in this tendency, which is common throughout the West, a mere wish to dignify the thing said. That wish is present, no doubt, but in a lesser degree than the determination to be comprehensive and detached.

Like the equation of science, the abstract noun takes its subject out of Time, where the verb would keep it: motives begin and end but motivation goes on forever. And once applied to a corner of the mind, the abstracting principle leaves no particulars alone. For example, business firms and other institutions no longer build a building, they put up a "facility."* This means that it will be fully equipped and serve several purposes. Then no one can object on

* The tendency to make a singular out of a plural collective noun—a tactic, a statistic, an ethic—looks at first like a reversal of the abstracting habit: the speaker wants a concrete particular. But this is not so; he wants the abstract umbrella still shading his particular. Otherwise he would say this move, this number, this code.

the score of fact: "You *said* an office building, but I see it has a restaurant and a swimming pool." The two tendencies of naming completely and doing it by abstraction are not opposites: they both give "objectivity" to the parts of life which we isolate for endless analysis, and they submit the result to community inspection, as science does with its objects.

The result for language is that it approaches a system of interchangeable parts—a machine. With the right terms in -tion and the usual linking words, there is no need to choose and group words with a care for etymology, sound, rhythm, or tone. Gradually, idiomatic expressions fall apart and syntax becomes unnecessary. The cohesiveness of language, so to speak, has been declining in all the western tongues, especially English. Foreign speakers have unintentionally helped to break the forms, but they have done less harm than such inventions as Basic English and structural linguistics. For both these systems view words as counters, to be assembled as if there were no constituted language behind them, no other speakers, but only an algebra of a + b in which terms are coupled to force a meaning by juxtaposition.* Basic is thus a typical artifact of techne and also a scientific abstraction that professes to convey the "essentials" by means of some nouns and a few weak verbs.

That by allowing this the method of Basic does not prepare the learner for the language—at least not until the language has wholly lost its set ways—is perhaps unimportant when the native speakers of English are busy getting rid of these ways from within. They make new words, at need or not; they telescope old ones meaninglessly, twist meanings, ignore, then forget, distinctions,

---

* "Even if we dispensed with case and person altogether, our sentences would make perfect grammatical sense: 'I gave *she* the book . . . I *is* here . . . You *is* here . . .' Their meanings are completely clear." These encouragements to the young go well with the introductory remark: "From the standpoint of speech, therefore, a sentence could be scientifically defined as *any stretch of utterance between breath intakes*." Harold Whitehall, *Structural Essentials of English*, New York 1956, 108, 29.

The counterpart of this is the "scientific" definition of reading: "a processing skill of symbolic reasoning sustained by the interfacilitation of an intricate hierarchy of substrata factors that have been mobilized as a psychological working system and pressed into service in accordance with the purpose of the reader." Quoted in *Council for Basic Education Bulletin*, September 1960, 11.

confuse persons and things (as in the false possessive: "Florida's governor") and grow deaf to connotations and cacophony. Some of this fiddling is nervousness; it works off the irritability of people who are held by their needs and occupations to a very exact memory of numbers, technicalities, and trade names. Language is the one malleable stuff they have left to play with, and play is liberating and amusing: it is creativity.

Besides, the city dweller under techne has the excuse that the speech at his disposal for the experiences of daily life is impoverished as compared with, say, that of a sailor or an old-fashioned peasant. Their words are more numerous, they are sharp, varied, fit, and pleasing in themselves. Even for the humblest they color the web of life.* Drabness certainly accounts for the spastic efforts of certain weeklies to liven up their prose by anti-grammar, pedantry, and other affectations. Most journals that are specialized in subject matter, yet must appeal to readers who are not professionals—for example, the excellent magazine *Trains*—develop their own "crisp" jargon, which is only to a small extent technical. The rest is contrived for color like the sports writer's prose; it helps make the readers feel knowing and apart. And besides, the impression is given that the writing is "efficient."

Efficiency is the modern-sounding ideal substituted by the linguists for correctness, and it is of course just as difficult to ascertain and apply. At one end, efficiency seems to mean giving the gist, the drift or minimum of information; at the other, it means the strong, striking, unusual way of putting things. Either extreme reflects the fact that language has become what the science-inspired linguists thought it was: a tool of communication. The possibility that language conveys meaning by intangible, nonmechanical means, by an inaudible surplus of meaning above the meaning of words, and which arises from their choice and arrangement—this is missing from the deliverances of linguists, as it is from the thoughts of a people schooled to think that a language is a piece of private, instead of public, property.

Lexicographers ratify the dogma that there is no right or wrong

---

* In comparing this quality in different cultures, we must be aware that the difference between *Metrecal* and *Liebfraumilch* is not in the substances alone.

way to speak or write, but only speechways—speech facts—which are all equally acceptable so long as the speaker's meaning is communicated. How this is achieved and tested and why in another part of society other experts are agitated by the aforementioned problem of communication is not explained. But one result of the combined dogma and problem is that when disagreement occurs, in business or sociability, everybody is ready with the platitude that it is a "semantic problem"—two things for one word. For the semanticists have popularized the belief that words betray, that one should always define one's terms, and that language rich in overtones tyrannizes the mind through the emotions. This "conditioning" occurs, no doubt, but it need not. There is an art of words which gives exact meaning without defining and without tyranny. But who can rely on it when the linguistic powers are misled and misused? To consider language a "tool," that is, to regard words as lying outside consciousness, ready to be used like a pair of tongs for picking up meanings as needed, is of course in keeping with the "objective" view of behavior. To counteract this, professional writers proffer the remedy we have so often met in techne: when two things that belong together have been separated, someone makes good the loss by supplying an "additive" to the deprived. The empty clangor of abstraction in our speech leads to the journalist's slick phrase and clever neologism, which are imitated in talk from the prose they enliven; an unjust war is waged on harmless old clichés, and violence is practiced on idioms to refurbish them. Soon, playing with language becomes compulsive. The urge is to surprise and be creative in the manner of a poet. And hence the excess of metaphor indulged in, even at the cost of making one's meaning obscure or forgetting it in nonsense.* Metaphor affects vocabulary: *pinpoint, spearhead, highlight; bottleneck, framework, floor and ceiling;* and it inspires the repetitious imagery of the executive routine—"play it by ear" (which is a misnomer for "improvise"), "wear two hats," and other circumlocutions.

Metaphor has also become the standard medium of criticism, social philosophy, education, and political debate, and too often of legal opinions. Throughout, the hope is to recapture the unique

* E.g., "This world-wide bottleneck must be reduced."

life, the colorful misbehavior of each act.* It almost always turns out to be merely bad poetry: the author of the manual on organizing conferences spoke of them as offensives and he talked of tactics when he meant catering.

This figurative habit is endemic in the scientific culture, for it serves the double desire of avoiding abstraction and of showing to the eye, without whose witness reality nowadays seems dim. Yet these spurs to imagination destroy it by excess, while enfeebling the words thus abused. From the ease of putting abstractions end to end and letting images do the work of thought, the reader contracts the habit of never thinking what particulars properly fit under either and he grows more and more detached from experience. Formerly, good prose was that which struck a balance between abstract and concrete and reserved the effect of imagery for explanation or emphasis. The reader was not put to work visualizing strings of metaphors or thinking up instances to fill out abstractions. Prose could be exact without being either flowery or dull: it was not trying to be either poetry or science.

While this transformation of language was going on at large, poets and novelists also were experimenting (as they called it) in an effort to restore freshness and force to vocabulary and syntax. The writers of the past hundred years, from Rimbaud and Mallarmé to Joyce and Faulkner, have had to play with language in deadly earnest, struggling to keep meanings sharp and sensations uppermost, because the forces of literacy were making words smooth and empty.† But the liberties writers took in protest only removed another barrier and scruple. Everybody's urge to innovate, to be free and creative, seemed warranted by the work of Hopkins and Yeats and their progeny. By now the punning of Joyce, the alliteration of Hopkins, and the grammatical freedom of Gertrude Stein,

* Kipling was a pioneer in this "effective" sort of description. His old steamers always "waddle" out of the harbor. The device becomes a systematic impropriety in terminology, as may be seen in almost any novel: ". . . the wind nudging the ship along, boosting it by the stern . . . cuffing the port side. . . ."

† "The English language has, in fact, so contracted to our own littleness that it is no longer possible to make a good book out of words alone. A writer must concentrate on his vocabulary. . . . He must let his omissions suggest that which the language can no longer accomplish." Cyril Connolly, *The Unquiet Grave,* 1945, 98.

Faulkner, and Proust are in the common domain of conversation, advertising, art criticism, and industrial technology. This was to be expected in a unified culture in which poets, advertising men, scholars, journalists, artists, critics, and politicians* all have a stake in working their will on the language, and are encouraged by the scientific linguists to believe that here is play that has no rules and tampering that is free of penalties.

The reader who knows that nine tenths of our old-established words are originally metaphors may wonder why I single out that verbal device as a chief evil in our speech, from which it spreads its infection to our minds. The answer is that there are metaphors and metaphors. Those embedded in the language possess varying degrees of energy, ranging from dead to live, and the dead metaphors form no part of the excess I observe. When one says *examine*, no one thinks of the pointer of a balance. But differently put, the same image (as in *weighing* an opinion) still has life and must be used with caution: one should not weigh an opinion to see if it will hold water and stand the acid test of public scrutiny.†

But this muddled metaphoric style, common as it is today, is less a condition than a by-product of the intent I see giving shape to our sensibility and, most lately, affecting science itself. The principle of metaphor is the joining of distinct ideas to show their likeness and suggest also their difference. Now, man's new power to transform matter and combine processes makes him relish above all else the idea of double natures and "two-way utility," of three-in-one substances and agents made up of miraculously fused incompatibles. The invented names for these things are in effect metaphors. They are also puns and portmanteaus in the manner of Lewis Carroll and James Joyce: *aqualung, transistor* (transfer + resistor). Even in the new words for a single idea there is often a duplicity in the echo-

---

* The politics of Marxism, which led to the renaming of familiar facts in a tendentious and pseudo-technical way (deviationism, agrarianism, bourgeois moralism, negativism, etc.) have greatly contributed to the decay.

† *Acid test, crucial* (experiment), *psychological moment* and many other metaphorical uses of phrases that the public connects with techne and science began to enter common speech seventy-five years ago, yet few if any added precision to the workaday language. *Focus* was borrowed from photography by the educationists half a century ago and we have not had a clear statement from them since.

ing of others that possess dignity and force. This is so, for example, in *automation* and in *microbus* and all the neologisms in *-matic* and *-rama*. When a megaton bomb was being talked about, gas stations put up signs about their product's megatane power. The same urge to borrow and enhance moved the writer who proposed to replace *weightlessness* in space flight by *abweight*.*[3]

To be sure, the mixing of Latin and Greek and of either with English to supply new names did not begin with our century, but the practice was bound to deteriorate when those playing with classical roots did not know the language they came from or care for the one which they were enlarging. In truth, there never was an age so poor at naming things. One has only to mention all the *-tron* hybrids which are the acme of meaninglessness.† But there is worse in the growing list of cases in which confusion seems to be deliberately courted. It is foolish to call projectiles Jupiter and Saturn when two of the planets already bear those names; it is downright stupid to choose *plasma* to mean an ionized gas at high temperature, when plasma already means the fluid part of blood and a kind of quartz. *Plastic*, which used to mean malleable, shapable, is now associated with hard grainless materials (and in Europe with portable bombs) so that *plasma* and *plastic* now evoke simultaneously the ideas of hard and soft, of blood and art and explosives, of solid and viscous and gaseous and fluid.‡ Until this triumph of nomenclature we knew that anything could mean anything; now we must wonder whether something can mean everything.

No doubt those who use these modern ambiguities are specialists who will not cross the line where each word begins to mean some-

---

* It may be too much to say that there is a germ of metaphor in the acronyms with which our minds are cluttered, but word-play and double meaning are present: the original title, which is made up of long abstract nouns, is designed to provide initials that spell a word such as SHAPE or SHAEF. It thus serves as the poor man's scientific symbolism.

† These compounds are obviously made by persons who think that the second half of *electron* (Greek for *amber*) has some meaning of its own that can supplement the various roots to which it is soldered. This collective ignorance and tastelessness have given *cyclotron*, *cavitron*, *climatron* and *phytotron*, all created by scientists. *Accutron*, *Puritron*, *Insectron*, *Chemetron*, and *Post-tronic* (for posting books in a bank) doubtless came out of the sales conference and public relations laboratory.

‡ There is the further complication that the study of the gas—plasma physics—divides into the study of stream plasma and discontinuous plasma.

thing else, so that no physical danger is likely to result from the equivocal sounds. Yet the duplicity, at once careless and deliberate, is harmful in a way more pervasive and permanent than any material mistake. The contemporary world is hard enough to understand, its parts are sufficiently strange and its meanings cause enough fright, to make all possible clarity desirable. It would seem but elementary caution in a scientific culture to call things by their proper names. But the tendency is all the other way. One does not know where to turn to avoid misnomers. In the excitement of war, press and public speak of the *rape* of a small country when they mean its seizure by armed force. Afterwards they speak of material deprivation as *austerity*, when there is nothing austere about either those who grumble or those who keep cheerful. If prominent American families engage in public life, they are at once called "dynasties"—clearly with no heed to what a dynasty implies. This mania for false analogy is now so usual that the startling sobriquet no longer produces its intended effect. Yet it continues to blur mental images. When a swamp near New York was made fit for school children to go there and study living things, the newspapers felt compelled to call it a "classroom," though one would have thought that a swamp was precisely *not* a classroom, but its opposite and counterpart.

If one passes from newspapers to scholarly prose, say, that of the behavioral scientists, one finds the misuse of terms no less frequent. The very word "behavior" is applied, again by a kind of metaphor, to entities that do not in fact behave. In economics, for instance, the phrase "cost behavior" means nothing more than variation. Nor are the humanists more sensitive to ambiguity. Philosophers of science speak of "lawful behavior" to distinguish the movements of physical objects from the actions of men, "lawful" being here a sort of pun like "durable" when applied to protracted lawsuits.

Indeed, punning, which has been barred from conversation and literature as contemptible (except when it is profound) has returned as the serious business of experts in communication.* Play-

* What other than a professional excuse could there be for choosing "May Day" as the convenient phrase to speak out over the radio as the ship is about to sink? If the origin is *m'aider*, as is commonly thought, it is inefficient, for no French

ing on words is of course a great device in advertising, where the prohibition of common words as trade names has engendered a second vocabulary of misspelled homonyms—*eesy, rite, expaditer, nutrament*—and innumerable portmanteau creations (e.g., "leisuals" for informal footwear), which may not last long in current use, yet which make their mark upon the eye and mind. Children's difficulties with spelling and composition are surely not lessened by this propaganda for wrong vowels and substandard words.

All this suggests that a part at least of the continual slithering of meaning in modern tongues is attributable, not to ignorance or carelessness by themselves, but to a deep sense of the instability of things. This feeling finds vent in the hybrid words (e.g., *Stellerator*); it informs the cant metaphors (e.g., "population explosion"); it shifts precise technicalities to vague uses (e.g., "epidemic proportions" to mean "very large"); it looks restlessly for variants (*difference* becomes *differential; basic,* basal; *advice* from the weather bureau is given in *advisories*—surmises will no doubt be called *suppositories*); it reduces a whole range of words to the service of one old idea (e.g., *pecuniary* becomes *financial* which becomes *fiscal,* until one hears that Mr. X, who is short of cash, "is having fiscal troubles"); it feels the need to translate the simple and personal into the lofty and general, often with a confusing idea added (e.g., the *poor* are the *underprivileged,* which may be so, but why privilege, suddenly, in our democratic life?).

This instability that I detect explains also the desire to fix the fluid world by compounding terms in the manner of chemistry: the "power-order conceptual framework" is as near as you can get to "trinitrotoluene" in political science. Tumbling from idea to idea through fugitive association lends the appearance of knowledge and mastery, although underneath there is a culpable absent-mindedness. Thus the engineer directing the Venus Project is called "the present-day Jules Verne"; and when a cherished film star commits suicide, a television retrospect of her career is entitled

---

ear will recognize any native verb in the sounds of the English phrase, and conversely no other speakers will easily catch the closed vowels of the French pronunciation. As for the overtones of the word "May"—maypole, Queen of the May, the merry month—none of these, apparently, should cause the men of the merchant marine to resent as an irony the new signal for their last day of life.

"Who Killed Marilyn Monroe?" What Jules Verne actually did or why it is absurd to represent Miss Monroe's death as murder by society are questions beneath the notice of our sophisticated misusers of words.

Nevertheless, it is shortsighted to suppose that this large-scale garbling we permit and come to relish does not alter the strength and quality of the language.* Some vagaries disappear but their traces remain as forms of thought, and it is only the unperceiving linguists aping science who think that a belief in the aesthetic virtues of language is an illusion fostered by writers. "Words are the people's; yet there is a choice of them to be made," as Ben Jonson remarked.[4] When the principle of choice is destroyed by the combined efforts of ignorance, trade, science, Basic English, and machine "translation," the language will no more be recoverable from the fragments than the Taj Mahal would be if ground to powder. English has a fair chance of becoming a world language if something of its spirit keeps it a language, but the increasing use of it as an inelegant algebra must bring it to a condition of which the sympathetic observer can only say: "They have neither Speech nor Language but their voices are Heard."

If what I discern is in truth a powerful tendency, someone may object, "Why argue against a trend?" To which the answer is: "What difference between yielding to a trend and giving in to conformity?" Besides, a trend is variable and may be reversed; as history, which is the graveyard of trends and the birthplace of countertrends, amply shows. I go on therefore to discuss the inroads of metaphor and duplicity into science. The breach was made when, as we saw in Chapter V, contemporary science acknowledged that its work consisted of regarding phenomena as—some convenient entity or model or system. Regarding-as is the essence of metaphor and ambiguity, and when this discovery seizes hold of the mind, it can come to seem a key that will open all locks. In a suggestive book of essays entitled On Knowing, the Harvard psychologist Jerome

---

* If the reader has noticed my use of "ologies" as an independent word, he may be interested to know that I am not innovating. J. A. Froude and perhaps others used it as early as the middle of the nineteenth century; for the word occurs in a letter by Thomas Hughes's father, dated Jan. 25, 1849: "Before steam and the 'ologies came in . . ." Thomas Hughes, Memoir of a Son and Brother, 1873, 106.

Bruner has developed the idea of metaphor to the point where he himself seems scarcely willing to write a literal sentence. His thoughtful work is full of remarks such as: "The psychologist . . . searches widely and metaphorically for his hunches. . . . If he is lucky, or if he has subtle psychological intuition, he will from time to time come up with hunches, combinatorial products of his metaphoric activity . . . he will go so far as to tame the metaphors that have produced the hunches. . . . It is my impression . . . that the forging of metaphoric hunch into testable hypothesis goes on all the time."[5]

One suspects after many more uses of the word that "metaphor" itself is used metaphorically and that in some of the uses it stands merely for "conception" or perhaps only "statement." At any rate, the game is in full swing and not a day passes without some branch of science electing to express its hypotheses or conclusions in language appropriate to something else. The geneticists have gained the attention they deserved by ringing the changes on the idea of a code buried in the chromosome molecule. It does not *act* or *behave*: it issues instructions, information, and the like: "DNA gives the replicating cell all the necessary information for perpetuating its features." Similarly, the attempt to manipulate the factors of heredity is called "genetic surgery." Recently also, the spectroscopists have taken to speaking of the "fingerprint" that a substance —say an amino acid—discloses to their apparatus and by which it is identified. At first the metaphoric word is put in quotation marks, soon it becomes the accepted, the only word for the procedure or its result. There lurks in this a pedagogic intention, but it leaves every subject filled with alien notions of doubtful clarity.*

To the extent that the popularization of science reaches a widening public, all this bad poetry spreads the sense of total interchangeability among things, and thereby strengthens the mechanical principle.

Certainly, the most dehumanizing of the current metaphors is the now deeply rooted one which calls electronic computers brains,

* The humanists are not slow to catch the trick: to justify the open-stage production of plays, they feel bound to say that it is "the most appropriate spatial metaphor for the public and esthetic questions of the day." Editorial in *Tulane Drama Review*, Winter 1963-64.

without quotation marks, and carries out this metaphor as far as it will go, which is to say from brain to mind. The machines are spoken of as learning and thinking and even "learning to make errors"; they are said to have memories, to simulate complex situations, to translate, and to teach. According to a circular from the Department of Commerce about translating Russian: "The machine is then asked to recognize the syntactical relations between the words." One waits for the day when it will return an ironic answer. Meanwhile it is not to their use but to the machines themselves as "collaborators" that results in textual study are ascribed; and for all I know they are thanked in prefaces. Some manufacturers, no longer content with successive models of such machines, refer to them officially as belonging to separate generations. Up to that point the public had merely been given the impression that computers afforded every kind of knowledge except carnal knowledge; now we are no longer sure.

Greater love of the machine cannot be conceived than that man should think his own mind inferior to the thing he has made. One hears it said in support of the brain analogy that "already the computer has outstripped the maker's own capabilities," as if this did not apply equally to a corkscrew or a pair of scissors. Man makes tools which work better for limited purposes than his own fingers. And he makes them by an activity of mind. It will be time to speak of computers as minds, or even brains, when they lay their heads or other parts together to make a man.*

But the machine, as sensible scientists have tried to point out, is not even a brain: "It is a remarkably fast and phenomenally accurate moron, rather than a 'giant brain.'" And again: "When an electronic engineer next tells you that a new electronic brain is as good as the brain of man, ask him when he expects to make a tube that will renew the metal of which it is made and not lose the specific 'memory' change [charge?] which it holds as a result of some specific past impulse."[6]

---

* The late John von Neumann, I am told, proved mathematically that a machine could be "programed" to engineer the construction of another machine. One need not question the proof in order to question both the feasibility and the implication of the proposal: so far, no machine will program itself.

One would suppose that practiced metaphor-makers would be sensitive to the differences between what they join; that they would remember, on the one hand, the gap between brain structure and consciousness; and on the other, the danger that what we are totally ignorant of should appear to be understood by recourse to verbalism. Such a self-discipline would not preclude admiration for the electronics engineer or wonder at the magic of his circuits, which do such useful things at superhuman speeds. But these speeds, like the accuracy of the performance, are a clear proof that we are not dealing with human minds. The people's good sense might have jibbed at the bad poetry, if the older and sounder view of language had still prevailed when computers first became known. But the mystery by which minds are read through sounds and by which ideas are transfixed through marks on paper was by then forgotten. The "mechanism" was thought to be understood; so all care was abandoned and misnaming pursued its frivolous fancy. Thus it comes about that people with Ph.D.'s who observe the ritual accuracy of their clan say they have a "translating machine" when they have mechanized a narrow vocabulary, and a "teaching machine" when they have mechanized a drill book. Had they invented the printing press they would be calling it a reading machine.

I do not deny the startling effect of modern computers on the impressionable senses. Seeing what they can do, one can in a playful mood imagine their turning into flesh and claiming cousinship. But scientists and engineers, we thought, are not licensed to be impressionable in this way. To excuse them one may recall that from the earliest history of computers, the analogy has hypnotized some minds, the first two on record being women. We owe the earliest computer to the mathematical genius of Pascal. In 1640, while helping his father, a tax assessor in Rouen, young Pascal conceived of a machine that would perform the tediously long additions and multiplications by making the figures on paper correspond to the movement of interlocked wheels. In the contemporary state of the practical arts the making of such a machine was enormously difficult. Pascal made fifty models and spent five years before the first standard calculator was made in 1652. It was this "Pascaline" that inspired the inventor's devoted sister Gilberte to say: "He reduced to mechanism a science which is wholly in the human mind."

Two hundred years later, Byron's daughter, Lady Lovelace, who, herself mathematical, assisted Babbage, the inventor of a notable computer in the 1840s, was struck by the same transmigration of intellect. The machine, she wrote, "is capable, under certain circumstances, of feeling about to discover which of two or more possible contingencies has occurred, and then shaping its future course accordingly."[7] In the next generation the logician Jevons, maker of a syllogistic machine, falls into the same anthropomorphic error when he writes: "Any question can then be asked of the machine, and an infallible answer will be obtained."[8] But although attenuated in words, the nonsense is as great, for "asking a machine a question" actually means pushing knobs, so as to bring into view symbolic letters which then must be interpreted to give the "infallible answer." The modern "program" on the most complicated machine in use does not differ from the principle of this sequence.

The harm of the metaphor lies in its reversal, which asserts that when one man asks another a question, what "really" occurs is a pushing of knobs and inferring of meanings from signals. We are continually "programing" one another and reading off the by no means infallible answers from the invisible tape streaming out between our lips. This may be excused as an instance of regarding-as —man *as* (not *is*) a machine. But the make-believe of that "as" is hard to hold in leash, particularly if behind the use there is a hidden animus. That a motive of hostility to man exists in modern thought is easier to feel than to prove. Certainly, much that is said about man, directed partly against those addressed and partly against the speaker himself, is an attempt to express the unimportance of his life. This emotion is linked with the masochism I mentioned in discussing behavioral science and with the joy of self-wounding that became popular in nineteenth-century accounts of the blind destructive cosmos. Our latest metaphors, shuttling between man-a-machine and machines-are-men, continue to mask these feelings in a mood of Samson-like revenge which is both anti-human and anti-scientific.

The great Darwinian metaphor of natural selection opened the floodgates to all the rest,* and made possible a scientific culture whose verbal style is in direct opposition to the traditional spirit of language: where language distinguishes and separates, the bad poetry of science unites; where language brings together like things, the names and also the workings of science and techne merge and confuse the disparate. And it is inevitable that when the ruling thought says: "Nothing is what it seems," the words leave a hint that any one thing may turn into another.

That this should be the age in which a school of philosophy has taken language as the sole source of light upon man's oldest problems, upon his duty and his destiny, is understandable but not reassuring. When literary critics study a text in the belief that it is made up of units of meaning akin to the hard-edged units of science and mathematics, which they can therefore count and classify like geologists working in rocks and fossils, they waste their time and blunt their sensibilities. The philosophers of language make the same mistake when they try to analyze the reality of fact or feeling by scanning the words of propositions in a vacuum. "Not a sentence or a word," as Whitehead said, "is independent of the circumstances under which it is uttered. True scholarly thought never forgets this truth." He adds that common sense neglects these circumstances when they are "irrelevant to the immediate purpose."[9] Between those limits of sensitivity on one side, and neglect on the other, intelligence of the word finds at each moment the right course. No rules exist for understanding, which is why language is the one device capable of recording the misbehavioral. The fringe of vagueness around words comes from their long associations and inspired uses within one language, as well as from the property a word has of reaching out toward ideas that it does not strictly imply. Scientific language, and especially number, is designed to do the opposite: it cuts off the aura which would impede demonstration. But in so doing it loses the sensuous immediacy resembling that of

* It projected the image of nature *ruthlessly* choosing. Had it been named "chance survival on the basis of accidentally adaptive variations," there would have been less excitement and a better debate; for no mindless series of acts can *select* anything, and the mindlessness—which was Darwin's point—would have been evident.

experience, the power to make the unfamiliar known, which words rightly put together will afford. That is how language mediates between minds over a wider range than that of conventional, defined, and static ideas. It permits, quite literally, mind reading.

When therefore the verbalist philosophers, following the lead of the great Wittgenstein, make language a primary object which, if carefully examined, will show what the world is really like, they assume in discourse a behavioral regularity that is contrary to fact. They also assume against all reason that the simpler the language the more truly it will mirror reality. It is interesting to know that one of the declared motives for this analysis of common sense and the commonplace is to ground philosophy in a realm free of the debris of natural and social science. Yet to seek a single ground of reality and to survey it by method and system are clearly derivatives of science and techne.

I add techne, not for the sake of reiterating a phrase possibly too familiar by now, but because the influence of the machine is what haunts all our deeds and misdeeds in language. To take apart and find the system in the parts argues a previous fabrication according to plan. But language was not made that way, nor man, nor the life he lives as a conscious being. It is significant that it should be a philosopher not of language, but of the law, who points out that "in the hand-made, as distinguished from the machine-made, product, the specialized skill of the workman gives us something infinitely more subtle than can be expressed in rules."[10] This subtle discipline of the word may seem an illusion: the users themselves admit there are no rules—it is continual anarchy and misbehavior. Quite so, but that discipline none the less governs one department of human affairs where man's own misbehavior is precisely the subject matter. I refer to the law. And I put the legal philosopher beside the literary critic and the language philosopher because their ways of work are those of language, from which they depart at their peril.

My skepticism of behavioral analysis in literature and philosophy does not mean that I think language should never be closely scanned; or that because language is irregular, writers may flout the claims of logic and precision. In the task of finding fit expression there *are* rules—and their presence does not contradict the truth that in framing and grasping ultimate meanings there are none. The same paradox holds in the law, and for the same reason, which our philosopher puts again as a contrast between the hand and the machine: "In law some situations call for the product of hands, not machines, for they involve . . . unique events, in which the special circumstances are significant. Every promissory note is like every other. . . . But no two cases of negligence have been alike or ever will be alike. . . . The certainty attained by mechanical application of fixed rules has always been illusory."[11]

The law today shares the general disrepute into which has fallen all that is intellect yet not science or not trying to turn into science. The law is in truth the most systematic and direct antithesis to science. As such it deserves more attention than it usually receives from educated people. Not only do the institutions we cherish depend on law and legal thought, but the ways of the law furnish an example that might strengthen the waning confidence in the reality of the unique, the concrete, and the nonstatistical.* Public concern about the law is particularly needed today when the conquering spirit of behavioral science is beginning to see in the criminal law a fresh theater of activity.

Perhaps the shortest way to mark the difference between the legal outlook and the scientific is to point to two characteristics of the English criminal law: first, it sets its hand against statistical living, for it prefers to see a guilty person escape to having an innocent wrongly punished. That is why it protects the accused and even the known criminal. This is not "reasonable" but it is civilized —if by civilization is meant the social recognition of individual rights. To put the whole matter as a contrast: the law cannot parallel the doctor's way of testing drugs, for the doctor gains his ends at the expense of a few—as few, he hopes, as possible, but still at the cost of sacrificing lives.

In the second place, the law excludes certain kinds of evidence.

* Thoughtful lawyers, from Justice Harlan F. Stone to Dean William C. Warren of the Columbia University Law School, have suggested the desirability of a college course in law for those not intending to become lawyers. The suggestion is apt and the reasons for it should seem as obvious as those which established college courses in science, economics, or fine arts.

Knowing that life is full of incommensurables, knowing the difficulty of so understanding a complex situation as to leave no reasonable doubt; knowing the impressionability of the human mind, even the experienced and the well-trained, the law will not allow all ascertained facts to be presented in court—for instance, the fact that the prisoner has been previously convicted of crime.

It makes no difference to my present contention that English, American, and Continental courts differ. The temper of the law is radically unlike that of science—though just as sharp and just as much entitled to its technicalities—and one hopes that these two modes of mind will remain distinct. Some lawyers, such as the late Clarence Darrow, echoing a stereotype, complain that legal practices have endured the same for hundreds of years, while science is progressing all the time. Why cannot science reform the law? To which an English theorist has justly replied: "The principles of criminal procedure are not the product of scientific observation, but embody a system of values. These values do not necessarily have to be changed with the march of knowledge of the material world. . . . The rule conferring upon an accused the right not to be questioned . . . may be a good or a bad rule, [it] has certainly not been made better or worse by the invention of printing or the aeroplane."[12]

Here, shaking his head, the behavioral scout enters, followed by technicians armed with lie detectors, truth serums, and a retinue of experts in all the sects of psychiatry, criminology, and social work. All look upon themselves as masters of some recondite knowledge and upon the law as crude empiricism flouting their science. They cannot see why their conclusions cannot be taken on faith (or on oath) so as to determine the truth in the modern scientific way— using *all* the evidence—instead of fumbling for it through a fog of objections, precedents, and exclusionary rules. A growing body of writers on criminology, mental health, and human engineering presses this argument and asks for revisions of the law which either would attach a scientific mentor to every court or would in effect substitute the conclusions of behavioral scientists for the judgment of judge and jury.[13] These experts, having free access to the prisoner for questioning, and knowing how to "administer batteries of tests," would in most cases be able to declare the truth.

There is about these proposals a brisk tone and an aseptic look which flatter our prejudices. But the promise of truth should not be taken at face value. Consider the so-called lie detector, which may serve as the symbol of the proposed procedure. It is not, to begin with, a lie detector—the term is (again) a poor metaphor to describe a polygraph, or recorder of variations in breathing, pulse, and blood pressure while the suspect answers questions. The user of the machine frames the questions, and his interpretation of the physiological readings determines whether the prisoner has lied and when.* The caution of the legal profession about admitting such "evidence" is understandable: the machine's report is a kind of hearsay in an unknown tongue. But—and this is a typical side-effect of techne—the mere existence of the machine interferes with justice; for the refusal to submit to the test is taken by public and jury as a presumption—not to say proof—of guilt: the "infallibility" of machinery, the false name "lie detector," overcome all but the most rational; so that in cases where newspaper publicity and community opinion can affect the trial and the verdict, the safeguards laboriously evolved during ages will fail, destroyed by the still presence of a machine in another room.

The chance of justice is but little better when the psychiatrist or other expert is brought in to find the truth by "understanding" the accused and gaining his confidence. The law rightly disbelieves confessions except when given spontaneously under certain well-defined circumstances of place and mood. These restrictions are nullified by expert inquiry. It is urged in favor of expert intervention that the traditional adversary system of trial sets up a debate, not an inquiry. True: debate is not scientific inquiry, yet it leads to new knowledge. Political democracy relies on debate for sifting rival versions of the truth and so does, ultimately, every form of learning, including science. Throughout, public scrutiny and the ethics of controversy

---

* My one vicarious experience with the machine and the expert in charge did little to increase my confidence in it. Some students were suspected of cheating in an examination; the parent of one of them requested that recourse be had to the "lie detector," which he hoped would exonerate the boy. The expert reported that the test showed no signs whatever of guilt, but that he was sure the student had cheated. Only, the boy had "managed to cut himself off from the real world," and could thus take the polygraph test without betraying himself.

must govern, in order to safeguard, the enterprise. What is alarming about psychiatric interrogation in the law is, first, that by definition it cannot be public; and second, that it cannot be restricted: the inquirer asks whatever he wants, in any way he chooses. The tradition of psychology and psychiatry and their avowed purpose is to probe to the depths and tell all, regardless of consequences: the truth shall make you free. The frequent lack of reticence among psychiatrists off duty shows a related desire to battle conventions and emancipate narrow minds. The law is far from being so naïve: it protects many kinds of privacy, knowing that under a regime of total disclosure social life would be impossible. As for the accused, innocent or guilty, they will return to live with their fellows and must not be stripped of all dignity in their neighbors' eyes or of self-respect in their own.*

A second objection is the unfairness and unreliability of the system of sympathy. With the spectacle of brainwashing, also expert, before us, it is evident that in conditions of isolation and fright a witness may say anything to gain the goodwill of his questioner or even more tangible benefits through the latter's intercession.† The adversary system, on the contrary, supplies a sense of partnership in vindication, which in a civilized society is a privilege of anyone on trial, even of the guilty. That is why "science" has no business in the direct determination of guilt and innocence: its unrestricted probing and its angular assumptions are admirable for making inanimate things "give up their secrets"; they can only harm and mislead when forced on men.

We know, moreover, that the temptation is strong among adapters of scientific method to reduce the misbehavior of the criminal not to the regularities of the law, but to those of the be-

---

* For a discussion of the pros and cons on this issue, see Manfred S. Guttmacher, *The Mind of the Murderer*, New York, 1960, chapters 18-23.

† See a recent case, *People* v. *Leyra* (302NY353, 99 N.E. 2nd, 553-557). The transcript of the recorded interview shows that the psychoanalyst told the defendant at the outset: "I'll tell you what the purpose of my talk to you is. I want to see if I can help you." To this the defendant answered: "Yes, Doctor." The doctor asked him about his sinus condition and the treatment he had had, and in the course of the interview said, "I'm your doctor." The transcript shows that on at least forty occasions the doctor in one way or another promised to help the defendant. . . . After this interview the accused confessed. Monrad G. Paulsen and Sanford H. Kadish, *Criminal Law and Its Processes*, Boston, 1962, 868-69.

havioral outlook. This amounts to unofficial law reform. Instead of the social classification of offenders, there is the scientific uniformity of "the criminal," "the" mind of the murderer—or rather, as many rigid definitions as there are conflicting experts.* The law has its warring opinions too, but the principle of judicial review, unknown to science and scholarship, brings about a convergence which, though imperfect, is intellectually impressive.

This is not to say that the best legal thinkers do not find flaws and contradictions. The existence in this country of a Model Penal Code and in England of Glanville Williams' critical and comprehensive *Criminal Law* testify to the need of reform at many points, possibly including the abolition of the jury. But what is present in legal thought and absent from behavioral is an awareness of *persons*. One speaks of *cases*, which sounds heartless, but the reading of even a few important decisions shows with what care, imagination, and punctilious rectitude the judges of the land, from the lowest courts to the highest, enter into the situations, motives, and rights of the persons who fall foul of the law. The literature of criminology, psychology, and psychiatry gives nothing to correspond. It is rather in this would-be science of social misbehavior that one finds the person reduced to the case, the item illustrative of general truth. Nothing could be more logical, since the effort is to arrive at "laws" and as far as possible explain any instance by a uniform set of antecedents.† Nor does one find, after the recital of these cases and their attribution to some causal "matrix," any dissenting opinion to take up the same points and show them in a different light.

I am not, of course, suggesting that the behavioral scientists in

---

* This has begun, in the usual style of behavioral tautology, with "laws of criminal behavior." The second of these laws states that "A criminal act is the sum of a person's criminalistic tendencies plus his total situation, divided by the amount of his resistance. This law can be put into a formula: $C = \dfrac{T + S}{R}$."

David Abramsen, *The Psychology of Crime*, New York, 1960, 37.

† The habit of genetic explanation (background, antecedents) is relatively recent and can be approximately dated: "We have learned lately to speak of men's 'antecedents'; the phrase is newly come up; and it is common to say that if we would know what a man really now is, we must know . . . what he has been and what he has done in time past." R. C. Trench, *English Past and Present* (1855), London, 1873, 283.

the branch that may be called applied are hard-hearted. On the contrary, they almost always take a maternal view of their "patient material." In criminology they want to fit the punishment to the offender, regardless of ostensible injustice.* This and the advocacy of various brands of "cure" for crime in effect cripple the law. For with the usual futurism of science, the new intention is held justified by its goodwill alone: the newspapers are full of assaults and murders committed by "cured" mental patients and criminals deemed rehabilitated. Thanks to this benevolence and to such other causes as the growth of bureaucratic law, the multiplication of technical offenses, the one-sided revulsion against capital punishment, and the "psychologizing" of social affairs, the antique idea of The Law is losing clearness and force. People in business, education and government enjoy playing at the law in home-made tribunals. Reformers of the true law develop economic interpretations and debunk judicial decisions—everywhere, the application of a crude social and psychological causality is undermining the power of the social compact and the responsibility of the individual.†

The public interest would suggest that when society grants science the right to study the criminal *as if* he were not responsible, it ought to grant law the right to regard him as if he were, and the consequences should be compared. The law can show how economical of energy its use is when it is surrounded with majesty: it strengthens internal commandments and rules by mere influence. It is government by idea, an idea which upholds and protects in everyone the unified responsible self, while also making the mind secure

* A massive attempt to organize and teach the criminal law as an incident in a large social situation, which includes the psychological make-up of the individual, has been made at the Yale Law School by Richard C. Donnelly, Joseph Goldstein, and Richard D. Schwartz. See their book, *Criminal Law* (Glencoe, 1963), and my review of it in the *Yale Law Journal* for November 1963.

† While a French jurist writes: "Responsibility, which is the foundation of the penal code, eludes scientific analysis and is thus a source of error and confusion. It must be abandoned at least in part and used only to the extent that it may serve in rehabilitation" (quoted in Jean Nocher, *En Direct Avec Vous*, Paris, 1960, 189); a group of American jurists open a discussion with the remark: "Here we seem to hear the voice of a friend saying: 'There is no such thing as domicil; it is a mere conception. There is no such thing as marriage; it is an abstraction . . . there is nothing at all but behavior-patterns.'" J. H. Beale *et al.*, "Marriage and the Domicil," *Harvard Law Review*, vol. 64, no. 4, February 1931, 203.

through the relative stability of legal decisions. Its rival, the scientific view of misbehavior, presupposes instead a conflict of forces that come together at a point X called John Doe. His responsibility does not exist and cannot be defined, for what does it mean to say: "Every person acts in accordance with his own emotional make-up. In this respect each person is an active partner to his own act."[14]

On this view the person is bound to regard his misdeeds as natural, logical, predictable, and society and the law as illogical and perverse in disallowing them.* If the result of the behavioral "cure" were greater happiness for all, there might be reason for abandoning the older conception of the law. But the outcome is otherwise. We know this from an earlier attempt at sympathy and latitude in the name of progressive education: no schooling was ever more arbitrary and coercive, for both pupils and teachers, than this permissive system in which instruction was to be hand-tailored and the growth of mind and will was to follow an unspecified blend of impulse and nudging.

The psychiatric interests supported progressive education as they now support the disablement of the law, and no one who has seen their handiwork in the one institution can wish to see it reproduced in the other. Its spirit already vitiates the "enlightened" treatment of employed persons, denying there also the principle of even-handed justice in favor of a patronizing kindness toward the "problem person." The heartless rigidity of an earlier industrialism is being replaced by the worst autocracy—that of professionalized kindness—and the differences among men are overlaid by the subtle inequality which consists in adults treating other adults as children.†

Justice depends on similarity of treatment, and to this end the

* In the theorizing upon such matters the modern fuzziness of language is rife, just as in the pleas of defendants the habit of metaphor is strong. As one of the men accused of conspiring to assassinate de Gaulle explained: "My so-called desertion was really only *a prolonged leave of absence*. [Italics added.] I had asked to be sent to Indo-China. I had nothing to do and time weighed heavily on me. At the same time I met a girl and I followed after her. They called that desertion." *New York Times*, Sept. 3, 1962. Compare a recent New York production of Sophocles' *Antigone*, which was designed to show that her violation of the law was simply a way of proving the law stupid.

† For the most lucid and favorable account of the system I deprecate, see Paul and Faith Pigors, *Case Method in Human Relations*, New York, 1961.

law classifies like a science. But its groupings and its tests are subject to an ultimate question which asks: "When will it be just to treat different cases as though they were the same?"[15] Not uniformity, nor again total anarchy, but equivalence. The impression that the "law is a ass" on principle, and that it enjoys forcing unlike situations into one mold is not borne out by the charges to the jury and the decisions of appellate courts. It is rather the public and the practical statistician who would just as soon as not convict a horse thief under a statute against sheep stealing: that seems the sensible way, but courts will not have it so. The virtue of functional subtlety, which is to say the capacity to draw sufficient distinctions, but no more, between similar acts and to re-create the complexion of mixed motives, lies altogether on the side of the legal minds. One has only to read the governing cases on Conspiracy, or on the meaning of Attempt, to discover the gap in conceptual and expressive power that separates a judge from a behavioral scientist.*

It was not to make a trivial play on words that the theme of misbehavior in this chapter brought together law, language, and literature. The link is in their natures. In working with crime, negligence, and all other nonrepeating incidents of social life, the courts produce a kind of parable from which to reason and move in later instances. The parallel is close with the work of the poet or the novelist. All these users of words seek to embody perceptions of the real in a way that extends man's unrepeatable experience. As Henry James once said: "Art is our flounderings shown." Without literalism, yet with particular detail, the verbal arts observe strict limits. In literature this is form; in law it is relevance. The court record and the legal issue define what may be spoken of: all else is ranting. Again, both arts take cognizance of imagination and illusion: in pleading self-defense to a charge of homicide, the danger or the necessity of that act of violence need not have been real or actual to justify the deed. Yet the sense of fact is ultimately as

* I mentioned earlier in passing the effort of such a scientist to predict, by machine, the rulings that the Supreme Court will make. "Work of this kind is in its infancy," says the proponent of judicial prophecy. "Computers can aid the lower courts in ascertaining . . . how the Supreme Court will view their decisions." *New York Times,* Aug. 8, 1962.

important to the jurist or the poet as it is to the scientist—perhaps more so, if one agrees that modern science is on the way to pure abstraction. Legal judgment, at any rate, is always "subject to any state of facts an accurate survey may disclose."

To anyone concerned with the freedom of the individual, the present unavowed conflict between the misbehavioral view of life and its many antagonists can come to seem the principal issue of the age. The confused emotions stirred by the struggle drive ethics and literature into false dilemmas, paralyze philosophy, and give religion a meretricious appeal as a higher ground from which to disadvantage the opposition. The issue itself is not clearly drawn, because, as we shall see, the prevailing idea of the mind is clouded with echoes of psychiatric analysis, behavioral physiology, and computer mechanics. As the hero says in a modern crime story: "It's complicated. I'm a cop, he's a psychiatrist—everybody's got methods."[16] The methods have this confusing effect, because their limits not being understood, they are not accommodated to one another and each seeks supremacy; and again because as scientific "tools" they tempt the user to deny ultimate difficulties and to promise knowledge, utopia, finality. In these circumstances it is deplorable that the natural remedy for inflexible method, which is art, should also have succumbed—and not altogether heroically—to the hegemony of the scientific.

X

The Treason of the Artist

THE educated public of the western nations takes it for granted that the two independent powers in the realm of mind are science and art. These alone, as I said before, are regarded as self-justifying activities, untainted by the common motives of utility and profit. And when anyone doubts that two great powers can ever share one realm in mutual independence, the answer suggests itself that art and science are only two forms of one creative effort. Thanks to such abstract statements honoring all the disinterested works of the mind, most people readily recognize in our civilization a double-headed leadership, not stopping to examine as we do here the scope and effects of the two activities they wish to see joined. Artists and critics of art are themselves not altogether clear about their relation to science, however much they may want to combat its influence; and when they succumb, as they often do, to an unconscious imitation of science, it becomes evident that one power only, and that one Science, dominates the culture.

Once this ambiguous battle is understood, all the desirable differences between art and science reappear in full force. Far from being alternatives (as they would be if parallel forms of one activity), art and science are complementary *because* opposite. Art, which fights a foolish battle with Utility, is in fact the antidote to Quantity and Abstraction. For art resembles life in being an organization of qualities: it may care about size and scale, never about quantity. It never seeks to reduce qualities—that is, sensations—to common units of homogeneous substance which it can then add up. Conversely, the qualities it displays are always felt as belonging to the work inherently and uniquely: the work is not simply a formula that sends us back to some storehouse in which the abstracted qualities are kept. In other words, the "additive" principle of techne which separates and rejoins has no business in art. Indeed, one of the sensations afforded by art is the anti-techne sensation of work, the sensation of struggle, drama, variability, imperfection. When we say a "work of art" we are truer to what we feel than when we say an "art object,"* for the first phrase records not our knowledge of the maker's effort but our seeing his hand in every part.

Unlike science again, art is never statistical—and thus uncertain about the individual—but exact and concrete and particular: not behavioral science but the law. The generality we may draw from the particulars of a particular work of art is our own affair; it is not the artist's and it is never provable by the work unless we distort it. Art reveals truth and gives form to experience; it is charged with suggestiveness; it imparts or renews energy; it seizes the indescribable and holds for memory the unseen gestures of the spirit; but it is not applicable information, knowledge transferable at will like a formula. For once again unlike science, it rejects the social discipline of method and common forms of expression. In science, if you understand the terms you cannot help seeing the point; in art you see only if you see. Mind and will have little to do with that kind of perception. Such is the way in which art is like life and science the antithesis of life.

In sum, as John Jay Chapman said, we cannot hope to know what art is. And this goes for the artist, who is drawn to mystery, not as an investigator and resolver, but as a rival artificer and remaker. Life is an enigma, says the Sphinx to Oedipus; Sophocles, passing by, fashions an orderly replica which is even more baffling because we grasp its terms almost perfectly. Today, obviously, the artistic impulse of keeping the mystery as it is has a difficult time in the face of science and education.† Nor are they wrong in wanting to explain. If man is to enjoy on earth the emotions proper to a native,

* This anglicized form of the French has lost the limitation of the original: *objet d'art* properly refers only to small, portable, and hence unimportant artifacts. The Parthenon is not an *objet d'art*.

† Salvador Dali was adopting the simple, modest, and right attitude of the artist when he replied about his interest in science: "For myself, le more complicate things for this reason like it le best today: le nuclear microphysics and mathematics because no understand, myself no understand nothing of these. Is *tremendous* attraction for understand something in this way." Interview on Paul Klee in *The Second Coming*, Preview issue, New York, Sept. 1960, 12.

provision must be made both for feeling at home and for contemplating mystery. The two are not incompatible; any great object of contemplation (science included) can give rise to each. The difficulty is to enlighten these emotions equally.

During the past hundred and fifty years the one great attempt to stop the dry rot of the soul produced by the conditions of life in the modern world has been the handiwork of the artists of the West. Not the humanistic scholars and gentlemen, not the critics, historians, philosophers, and students of society, but the artists have had the intuition, if not the total vision, of what we are now all able to describe. The artists have caught from one another the tragic message that something essential to human life was in mortal danger, and with singular unanimity they have uttered warnings, improvised defenses, administered what they hoped were antidotes. They themselves have been driven by the obsessive conviction that to live under the industrial order and remain a man in the ancestral sense was impossible. Something pervasive that makes the difference, not between civilized man and the savage, not between man and the animals, but between man and the robot, grows numb, ossifies, falls away like black mortified flesh when techne assails the senses and science dominates the mind.

Clinging to this perception, the artists of Europe and America have toiled and thrown away their lives to make their anguish intelligible and contagious. Beginning with the Romanticists at the turn of the century of "light," they broke with society in anger and violence, denouncing now one now another feature of the hateful trap. They marked their dissent by living on the margin of pauperism or crime, by escaping into the dreams of drink and drugs, or by fleeing the West altogether and finding some still primitive refuge across the seas.

Fugitives or not, they left their words and inventions to show us what they felt and with what devotion they labored to make us see and resist our peril. Their chief device—if one may generalize about so broad a movement—has been to divorce us from the world by breaking the established bonds between our ideas and our feelings. Whatever the man of good will expected from life or society or art itself, he must learn to repudiate in order to find truth and pleasure in something other or opposite. What society gives ready-made: family, duty, propriety, ambition, power, stability, religion—in short, the program of virtues and sins, the schedule of social cause and effect—these must through art come to be detested and their contraries seen as good.

To this end, common speech must be broken up and skillfully misusaged; visual images must defy the commonplace and revive dead feelings by shock; explanations and commentaries must bewilder—if possible, outrage—in the attempt to shake the narcotized and the indoctrinated into regaining the simple use of their senses. As the grip of techne and science got firmer hold of ever-larger masses, the treatment had to become more severe. It was no longer a narcotized but a petrified population that art must bring back to life.

The means used in our own day for this resuscitation are no doubt familiar, but they will bear recalling through the mention of a few of their creators. The artist, to begin with, is for us so important as the martyr and savior of a satanic world that in our time he has become the central figure in the public concern with art. We learned to acknowledge him in that tribal role by reading James and Proust. Joyce, after charting once again the course of the artist-redeemer, plunged us back into the elements. Near the opening of *Ulysses* we are confronted, through Leopold's reverie, with the body's simplest need, and Molly's closing meditation is intended to leave us in the stream of life through the fact and fantasy of lust.

More didactic still, D. H. Lawrence made his fictions carry the burden of his search for an entirely new society, in which the senses and the mind would work in ways totally opposed to ours; for in our present relations with one another—as his characters' destinies demonstrate—not a single habit or impulse is right. Everything is vitiated by the intellect, which in its fears and calculations corrupts the passions and destroys health, just as it corrupts truth and destroys mutual trust.

Similarly, though from another starting point, Gide bade us

develop a desire for the "gratuitous act," that is, an expression of the self quite spontaneous and useless, hence truthful and satisfying, because wholly unrationalized. To make the lesson vivid, such an act can be a motiveless murder, which paradoxically represents life by being inexplicable energy. The importance of no-meaning to all these desperate creators of symbols is evident. To them, all purposes and reasons seem tainted with technicity and science. Anything resembling a formulated idea they think doomed to become part of the traditional rubbish, for any idea can be related to some other idea, classified, dealt with, and thus robbed of its force.

But the animus of our artists was strong enough to keep their anti-idea ideas from losing force at the point of impact. Rather, the frontal attack on the bourgeoisie, on its morals, intellect, and governance, conquered the imagination of the ruling and thinking classes throughout Western civilization. Though these chastened leaders did not quit the world, they learned to live, emotionally at least, in that anti-world depicted in negative by the artist—and not merely the contemporary, but any artist of the past whose presentment gave signs of the same subversiveness. Whoever in his day or ours has been abnormal, perverse, antinomian, irreconcilable, maddened into rebellion, or as much as malicious and impudent, has received a place in the modern pantheon. Our masters have been Baudelaire and Rimbaud and other accursed poets, Kierkegaard and Kafka, Melville and Yeats and Eliot, Rilke and Lautréamont, who in their several ways enacted or foretold the history of the wasteland.

Their works also hinted how another life might be lived, but self-reformation through the printed word alone is difficult. The plastic artists had a better chance to be the revivifiers, and they took it. By direct appeal to the senses they too broke up the crust of associated ideas. The painters forced us first to see the world undone and reassembled in a harmonious jumble of the remote and the familiar, the unlikely and the ludicrous. Then they made us experience pure sensation—color, line, shape, space, relation-without-thought. We were to bathe again in the happy confusion of infancy, but so ordered that it would arouse pure pleasure and give rest to old emotions. Here again no-meaning was therapeutic, a stimulus to long-buried dreams, a key to repressed symbols, a passport to the forgotten freedom of myth-making.

The sculptors, meanwhile, liberated us by showing that suggestive shapes could be made in endless numbers without representing anything but our love of balance and solidity, our desire to remold the actual or to experience the lust of touch and sight upon indeterminate roundnesses and hollows. These artists proved that the catalogue of created things need never be complete, and we carried away in our mind's eye and even in our fleshier senses a voluptuous feeling akin to that evoked by the poets and novelists.

At first sight it might seem that the architects, also makers of shapes, were bent in the other direction and wanted to deny our senses any satisfaction. Yet they too were simplifiers who stripped the familiar world of clutter and conventional beauties and remade it clean and smooth for our imagination to play upon. The wide spaces, the polished surfaces, the bare masses hardly altered by fenestration, were meant to clear the horizons. If any idea remained it was that of grandeur, produced by great height and intentional emptiness. And when after forty years the very virtues of this international style came to be felt as dwarfing and depressing, the genius of architecture burst forth in a riot of new impulses ranging from abandon to whimsy. The goal was irregularity at any price, free form to dispel the thought of the machine, a more human scale, and the secret of apt ornament. The builder today, not content with defying Euclid and his right angles, deliberately stands a massive structure on frail pillars, a pyramid on its point, because he is still thinking of the needs of man's ego, and seeks to soothe it by an ingenious indifference to gravity.

Like the plastic artists, the musicians also have striven to recreate a young world of sensation, and like the architects they have subdued the powers of techne to serve their ends. The composers of the past half century first did the work of destroying set feelings by deserting the traditional modes and contriving the shocks of atonality; then, seizing upon the devices of electronics, they began to produce fresh sensations for the ear. In our midst they capture the raw materials of a new art of sound by recording the world's din and rearranging it under the name of *musique concrète*; and they also fashion sequences of sounds familiar and unheard of under the name of electronic music. This last effort is undoubtedly the most radical and promising of all the artistic resolves that our senses and minds shall be made new.

Meanwhile the so-called "serial" composers continue to use the classical sounds and instruments in the task of finding for music a new principle of organization. Whatever one may think of their results, their search has a general significance applicable to all the modern arts. For it is not enough to galvanize the moribund senses. Modern man will not be made whole by merely taking in the new impressions raw. Cleanse his feelings and imagination as much as can be done by the bizarre, refresh his eyes and ears with original sensations, and he will still need an order of perception and some boundaries of meaning before he can feel pleasure.

But order and meaning are still denied him and he is, as we know, far from any prospect of pleasure. The vast onslaught of art upon civilization which I have hastily sketched here has left him more racked by pain than comforted by the intermittent chances of feeling and dreaming. And no wonder, for the medication, however well-meant, has been harsh. In the very house of flattery, the theater, he has been bullied and insulted. On the printed page and on canvas, he has been ridiculed and made a fool of. With the stiffening of the dose, which grows more bitter each year, the cure implied if not promised in the time of Ibsen, Shaw, and Wells is no longer even thought of. There is no *vita nuova* in the squalor of Beckett, Sartre, and Genet.

This passage from the wasteland to the refuse heap might at first lead one to suppose that the artistic crusade to save mankind has failed through the crusaders' forgetting their purpose, or losing patience with those they meant to redeem. I do not believe such a surrender to be the cause, nor that any lapse of courage or fidelity has brought about the present situation, in which we see the artist at the end of his resources for shocking and flaying, and the public at the end of its hope of being saved. One may question the degree of failure and despair, but the salient fact about modern art is that it is no longer the permanent and public antidote it started out to be. This is not because art is neglected or its influence restricted to an elite. Never, on the contrary, has art been so widely spread and publicly revered. Like science it has become a diffused cultural force. Within the short span of a quarter century, thanks to techne, the art of the past has become an important kind of consumer goods. Meanwhile, avant-garde art has extended its sway as a quasi-religion charged with antisocial animus. These two species, counterparts of the two sciences, presumably fill our civilized needs: the horrors steadily depicted in the new art are to act as a mithridate against the horrors of actual life. The democratic peoples' concern with art, as consumers and as makers, is supposed to occupy the leisure created by the machine. The official recognition of the national artists is intended to help reduce their alienation. And the "tradition of the new" —to use Mr. Rosenberg's phrase[1]—the perpetual revolution in artistic style, is meant to keep us interested, stimulated, up to date, as required by our habits formed on industry and science.

It is these very habits that prevail over that simple taste and denature art, both traditional and avant-garde. The art of the past, as many critics have pointed out, is vulgarized or destroyed by being, as we say, "channeled" through the mass media. More important, the art of today, caught between two legitimate commandments, is coerced and twisted into what it never meant to be. The two commandments are: to mirror and to criticize. As mirror, contemporary art re-enacts the work of the machine and its effects, of which the chief are emotional tension and formal abstraction; intellectually art follows science in its various tendencies.*

Modern art has followed science also in making the eye prevail: it is painting, sculpture, and architecture that attract and represent us best. Sound comes second and the word last. Victor Hugo's

---

* Whereas at the end of the nineteenth century, Hardy regrets the scientific tidings ("human souls may find themselves in closer and closer harmony with external things wearing a somberness distasteful to our race when it was young": *The Return of the Native*, chap. 1), the writers of our time take the "fact" for granted, as when W. J. Turner says of the mystery of naming: "It is strange that a little mud/Should echo with sounds, syllables, and letters." *Seven Sciagraphic Poems*, London, 1929.

epigram about the effect of the printed book on the storied art of
the cathedrals—the first will kill the second—is on the verge of
being reversed, picture and number choking the word.

So much by way of generality about the duty that art discharges
as mirror of our state: I shall discuss particulars in a moment.
The second duty, that of criticizing, should (one would suppose)
make art avoid the echoes of techne and scientism. But this avoidance is coupled with imitation and it gives modern art the queasy
quality by which it may be known.

In rebelling against scientific culture, modern art has been not
merely violent but "abolitionist." Its prevailing tone is suspicious-
ness, hostility, and ill-natured parody;* and its target, which is
ostensibly society, is actually those facets of science and industry
that seem ugliest to the particular artist: number and system,
which are "inhuman"; bourgeois morality and the profit motive,
which are inhumane; science and the state, which are anti-human.
In the end these "enemies of promise" are taken to be mankind's
fate, the human condition, from which it follows that life itself
must be regarded—and shown—as a hateful fraud.

Moved, then, by contradictory duties, the modern artist finds
himself doing the work of dehumanizing he abhors. By mirroring,
as plastic artist, he shows man attenuated and anonymous—witness
the sculptures by Henry Moore and Gonzales; man broken up and
disfigured—witness the last of the quasi-representational paintings.
By building as modern architect, he overwhelms and belittles man
in gigantic upended coffins of steel. By composing as modern
musician, he destroys the inner rhythm and concord of feelings
through the insistent beat of harsh combinations of sounds, latterly copied on tape from the clangorous life of street or factory.
By writing as modern poet or novelist, he gives no quarter to
either intellect or sensibility, pitting them against each other in

* Consider, as a single example, the typically sulky mood of Erik Satie and
its expression in such a work as *Vexations*, which is marked "Repeat eight hundred
and forty times" as an obvious (not to say sophomoric) dissociation from the
customs of society as well as of musical art. The fitness of this angry expression
is shown by the fact that the piece was recently performed at full length by a
team of concatenated pianists through the evening and night of Sept. 9, 1963.
New York Times, Sept. 11, 1963.

words and scenes by turns wilfully flat and opaque and sordid and clinical—in short, reiterating to the point of nausea the proposition which thinking beholders no longer dispute: the life man has made for himself is not worth living.

The public, long since beaten into docility, accepts everything, but prefers no-meaning and no-shape as being less fatiguing. The artist complies, being through his superior sensibility directly in touch with whatever stirs around him. But in so doing he faces his extinction and that of art, for the reduction of forms to simple hints, the elimination of moral and other ideas, the breaking one by one of the bonds with nature and humanity, must end—as it has for the Action painters—in the denial that painting or any other art exists: the act of painting is a piece of behavior in a private biography, and there is nothing more to say about it.

The logic of this position results from the fact that the artist's will and his sensibility are at odds. He cannot help abolishing the will, thereby abolishing society a second time, in himself, if that were possible. The will was originally to make art and through it to defend the wholeness of man's life. But this is a time when wholeness is denied. Art has no religion, no pantheism, no man-proud society to make common cause with: the only currency of ideas is analytic, not synthetic. Forced to become the enemy-interpreter, the artist suffers (like the common man among the machines) in having to love and hate the ultimate stuff of his art. He, or others for him, call it alienation.

But if this predicament of the modern artist's is inevitable, where is the treason? In the way he has taken his fate:—in his pretensions and ambitions, his whoring after ideas beyond his grasp, his excessive verbalizing and the idiom chosen for it, his faddishness worse than the public's and his attitudinizing worse than the critics', his misplaced self-importance and his failure to assert his true importance; in short, his defects of character, not as artist but as artist-and-social-critic, which the accident of birth compelled him to be. The treason, as often happens, was a falling into temptation rather than a positive act.

I must add that there is, of course, no such person as the modern artist, so "he" has not done these things, has committed no treason. Any living artist can exonerate himself by pointing to his deeds. But the fact remains that if one takes a catholic look at modern art and listens to the declarations of its makers, one is forced to a double conclusion: art seeks to be an Object, pure and free, and its chief claim upon us is its "art" in the sense of technique and method. To be sure, despite much critical effort, opinion about these matters is confused. What does art as an independent object mean in the face of abstract and non-objective art? Clearly, "object" is to be understood in the scientific sense of a pure construction, independent of the beholder and his ordinary emotions and ideas. Each work of art, says the critic, is a world in itself, with its own laws; it answers to no human demands but to the necessities of its structure and materials. The realm of art is thus a universe like *the* universe, a collection of objects "out there," like the objects of science. Often, the effort to make the object pure makes it "non-communicative" in the same way that science avoids the human by departing from common sense. In both art and science, method substitutes for direct perception. Thus as abstraction increases in art, maker and critic are driven to method in order to orient themselves and their audience.

Some excuse for this shadowing of science can be found in certain intellectual discoveries of our century that legitimately concerned artists: the analysis of myth in Frazer's *Golden Bough* and the exploration of the unconscious by Freud. Einstein's Relativity and Planck's Quantum did not really matter to artists except as these new conceptions denied the old "copy theory" of science and the fixity of solid matter. While physics seemed to say "anything goes, provided your symbolism is consistent," psychoanalysis very certainly said: "what goes on in the conscious mind is a mere film on the deep turbid waters of life." And those waters, Frazer explained, had been charted in the myths and magic of early times.

This knowledge should have been liberating; instead, it made artists and readers into students. Their "science" stiffened myth into a code. Even ambiguities occupied "levels" and no sensation or belief was allowed to subsist unchallenged and unplaced. This conviction was at the root of the grievance against Romanticism, which had been so careless and unscientific. The modern desire for multiple meanings responded to the super-position of elements in our crowded life. But it also answered the requirement for a kind of algebra in art, by which several values will satisfy one equation. Hence the model poem or painting that we take pride in unriddling.

Naturally, making these foolproof constructions means denying inspiration and rebuking the improvised. Though the artist is to draw on his unconscious for the right linkage of his perceptions, he must consciously order his materials in great anguish and "go into council for every word." The artist is said to be moved by a rage for order. Every symbol must tell; there must be no loose or empty places or the whole will not be defensible: it must hold up like the formula for a system in physics.

As witness to the agency of science in leading to this view, one may take that exquisitely fine, that most re-readable and limpid poet, John Crowe Ransom, whose doctrine is that logic and even metaphysics must govern the best work. The rest is amateurish, as he professes to show in analyses as strict as those of a conveyancer scanning a deed.* And the reason is clear: "the poetry of the feelings is not one the critic is compelled to prefer" to the "poetry of knowledge," because "it taints us with subjectivity" and involves all poets "in a general disrepute, the innocent as well as the guilty, by comparison with those importunate pursuers, the scientists."[2] What the scientists pursue is named as "the object," which a "subjective" poetry fails to pursue.

No more explicit statement could be wished of the modern rivalry between art and science for the same object, and of the exclusion it compels art to make: man, the subject par excellence, without whom no experience is, must abolish himself in art as in science for the sake of a poetry of knowledge—a Baconian art that Bacon never imagined—justified by the consistency of symbols and the logical structure of ideas. The handling of words in a poem to

* See the famous essay "Shakespeare at Sonnets," in which we are told that an image may be "very fine for a metaphysical poet who will handle it technically—for Donne or another University poet. It is not fit for amateurs" such as Shakespeare, who "did not go to a university and develop his technical skill all at once." This accounts for Shakespeare's repeated failure to be "systematic." *The World's Body*, New York, 1938, 299, 272, 286, and *passim*.

achieve these ends is specifically "the problem of poetic strategy."*

In other words, starting from the legitimate claim that since the decay of religion art alone has possessed the life-giving revelation, modern artists have succumbed to competing with science and taken on the characteristics of the enemy they sought to destroy. Therein lies the treason. At first, the desire to interpret art as happening from some lofty necessity was merely a manner of speaking; it was modest and boastful, as when Auden said that the poet must haunt the places where words carry on their own dialogue. Poets and other artists knew very well that they are men and that their toil with words or paint does not occur like a natural process to be watched. But soon the seepage of science through the culture forced a shriller note and more histrionic attitudes. A painter like Braque declared that he "did not believe in things, but in the relations among things."[3] This is science. Le Corbusier, contrariwise, was moved to utterance by techne: he thought a house was a machine for living and that the greatest modern artists were our bridge builders. Architects generally began to disown their ancient connection with art and to assert their true kinship with engineering and sociology: "all the architecture can blow away and leave a fine building there."† In the end art is not art but the analysis and abstraction familiar to science: "Men do not build houses with materials. They merely organize visible-module structures, comprised [sic] of subvisible module-structures."[4]

Painters, meanwhile, indulge themselves in the hope that "white lines in movement symbolize light as a unifying idea which flows through compartmented units of life bringing a dynamic to men's minds ever expanding their energies toward a higher relativity."[5] This is not the pathetic verbal effort of some solitary and misguided artist: it is the standard speech of the guild. The creators of pure design from Mondrian to Albers set themselves "problems of line and space" and called the "solutions" by such names as "Explora-

---

* Again according to Ransom, Shakespeare is a careless workman, who "never had to undergo the torment of that terrible problem, the problem of poetic strategy." Op. cit., 278, 272.

† Heard at a meeting (1962) attended by twelve leading architects, where it was also said without anyone objecting that "architecture has become the first of the behavioral sciences."

tion III"; or, like the late Antoine Pevsner, they leave at their death a two-ton bronze entitled "Dynamic Project in the Third and Fourth Dimension."[6] Stronger minds, faced with the task of pedagogy, as Paul Klee was in the Bauhaus, have given support by works like his monumental *Das bildnerische Denken* to the idea that there was not simply a technique of art to be learned and remade from time to time, as artists have always done; but rather that art now was an exact science of reality, whose exposition required as many graphs, proofs, and chains of demonstration as a treatise on astrophysics.

Musicians, like many painters and some poets, have been caught by the same desire to be working at science, in keeping with which they have named their works as if they were reports of experiments, blueprints of machines. John Cage's scores are known as graphs, and his explanations tend toward loose metaphor, drawn from admired realms other than art.* The common use of numbers after each version of a work completes the impression of a technical feat, just as in every art and its criticism the vogue of the word "dimension" (to mean aspect or element) suggests the possibility of measurement. Indeed, from the outset "experimental" art has shuttled from the "workshop" to the "laboratory," and furnished words, ideas, and attitudes snatched from science and techne.

All this might still have been written off as superfluous wordage if evidence were not at hand that it represents rooted belief. The evidence is the flourishing of various "arts of factuality." Beginning with the Surrealists, painstaking work in all genres has been done— or rather, recorded—under the influence of hypnotic automatism or of drugs; and this not like the Symbolists, from helplessness, but like scientists, from heuristic hope. We thus have had the poetry of delirium unedited. Similarly to deliver the thing-in-itself, a school of French painters thrusts the object at the beholder before it is

---

* See his *Lectures* (New York, 1961), in which we are told that "We're not going to become less scientific, but more scientific. We do include probability in science." Mr. Cage prefers to call the composer an "organizer of sound"; composition is a process, it is experimental, and the organizer of sound is "faced not only with the entire field of sound but also with the entire field of time." The music "exists without psychological impulse" and "mathematical precision replaces the capricious meanderings of human methods." See Paul Collaer, *A History of Modern Music*, New York, 1961, *passim*.

"processed." This is called *l'art brut*, the natural datum, so to say, after it has been found by "exploration"; just as *musique concrète* offers the noises and rhythms of life untouched though interlinked. "Theater-in-the-round" is a milder form of the same passion for the inartificial.

For all these the public was prepared by its earlier acceptance of the chunk of machined steel on the mantelpiece and the photograph of hormones hanging above it: art straight out of life. In this *art brut* or *informel*, the model is techne; "reality" is given out in fragments by analysis or "research." But there is another path, which is abstract in the way of pure science. Instead of physical objects it offers the products of a conceptual scheme. The twelve-tone row in music seeks form, like all art, but does this through a regulatory system, "structuring," arbitrary, born of theory; and though affording full play to an ingenuity for the eye, disregarding the ear as science disregards ordinary objects.

In painting, system based on playing with numbers has produced designs corresponding to the measurements of external relations arbitrarily chosen. The numbers gave the length and defined the angles of the lines making up this "rigorously conditioned work of art." At the same time, notions are derived from the statistical character of scientific laws. Art too must have "randomness." Accordingly, scattered dots on paper may give the start to a musical composition, or lists of words revolving on drums supply a "poetry machine."

Handwork was kept for throwing or dripping paint on canvas. In each chance performance the "results" are obtained without the interference of man and his "subjective" mind. No longer an activity for our most acute and daring artists, art is now a process, and as such fittingly mechanized: man's proper role is that of a crank.*

As the most talkative and ubiquitous of the arts, prose fiction could have been consulted first on the change in our artistic beliefs,

---

* In the spring of 1963, Picasso announced that he had perfected a painting machine on which he had been working for some years. London *Times*, May 1, 1963. In making machines for a satirical purpose, others have imitated Tinguely rather than Picasso, who sought to "shorten the time between what the brain conceives and the hand executes."

but the demonstration might have been less convincing: it has long been true that the novel worships fact and that it has often tried to supply the lack of a behavioral science. Yet there are degrees in factuality and novelistic sociology, and what has happened within the span under review shows that the novel has not avoided the treasonable flight from art.

In the classical novel from Scott and Balzac to James and Zola, fact was sought for its suggestiveness. It lent weight and verisimilitude but was freely handled even by the extreme naturalists.* It was Proust and Joyce who first responded not merely to the sense of fact without which the novelist goes astray, but to Ritual Accuracy, which is the homage art pays to scientific research. We know how Proust cross-examined his friends about details beyond his reach, just as we know that Joyce could not rest easy until every indication in his *Dubliners* and his novels had been verified as to place and name and distance. And now, thanks to an academic critic who is also a superb photographer, Joyce's admirers can buy a book of Dublin photographs showing the main scenes of the (partial) fictions.[7] Readers expect precision or they withhold their assent.

This preoccupation has favored the rise of a school of writers whom I once permitted myself to call "nonfiction novelists." Gide was among the earliest; many are still living and producing works in which the interest lies in the care with which the "study" has been conducted. Though characters move through depicted scenes, little or no effort is made to render life; the presentment is an abstraction, like Picasso's thin-line bull. But every detail is exact, every statement searching: they can be tested by reference to someone's psychology—Freud's or Jung's—or someone's sociology or mythology or symbol system. The material is from the best clinic, and there is in truth no reason why it should not take the form of a report. Dullness is a main attraction, as we find in such works as Vladimir Nabokov's *Pale Fire* (which is a fiction in the guise of a piece of purposely dreary scholarship) and in other fictional retreats from individual experience: the modern novel is really a sounding

* Compare, for example, Zola's *Germinal* with the notes he collected from factual sources during his preparation.

into the collective unconscious masquerading as characters, feelings, and situations. Man counts for nothing in his actions.

This philosophy is unavoidable, for when the language is taken from the textbook there comes with it the tacit denial of spirit. It is not only in novels about doctors that one finds chapters entitled "The Fatigue of the Synapses";[8] and it is to suggest something like Tennyson's "idle tears" that the novelist writes: "When his nervous system and his tear ducts made him cry. . . ,"[9] though the result is to imply a mechanical act and attribute it to a material cause.* In the poets, scientific jargon is for "irony," in the novelists for verisimilitude and that necessary knowingness of the narrator; but wherever found it ends by making all tales and visions documentary. The logical next step, which is indeed the latest invention of poetic mind, is the composition of a "happening"—a series of indubitable events, unrelated to before and after, sometimes comic and always aimed at showing the incoherent absurdity of life. The "happening" as a terminus for literature parallels the machine- and the drip-painting, the random-point and random-word poem, and other methods for securing art without artists—which indeed the bourgeois public already does by holding exhibitions of "art" by children, convicts, primitive grandmothers, talented chimpanzees, and the sea waves acting on beach debris.[10]

If in these conditions anyone wonders why there is before us no model of man; why the hero has disappeared from literature; and why the tragic sense of life, which is exhilarating, has been replaced by the sense of an absurd life, which is debilitating, the reply must be that *anthropos* has been subdued not only by the material forces of industry, numbers, and the rest, not only by Marxist ideology and the babel of upper-mass culture, but also by the sincere artist's capitulation to scientism. Swinging from parody to imitation, modern art has left as the sole object for

---

* Down to the middle of the nineteenth century, after the first wave of mechanism in common thought which followed the work of Newton and Locke (though neither was a mechanist), the poets and writers generally showed themselves much more resistant to the doctrine, though receptive to the new ideas. As Marjorie Nicolson has lucidly set forth in *Newton Demands the Muse*, from Pope through Thomson to Blake, the vision remained of a self-animated world in which color and sound and scent were real properties. Princeton, 1946, 135, 143, 157, 236-37.

man's contemplation and comfort the vision of a life without conscious origin or purpose, a machine without a machinist.

Before our age attained to the accidental in purposeful design, our artists had produced a large body of more calculated works which answered, as we saw, the needs of mirroring and criticizing the scientific culture. Most of these works were elaborate, subtle, opaque—in a word, difficult. And the public was warned that the difficulty was deliberate, a challenge to understanding due to a high concentration of intent in the object, which was not a message but "a world in itself."

To this object, this structure methodically made, the reader was expected to bring the appropriate decoding equipment, and since that was perhaps asking a good deal, the critic stepped in with *his* method. Considering how unanimously the moderns in all the arts have represented their work as rigorous in conception and construction, the possibility of an equally rigorous method of interpretation seems reasonable. And obviously it is needed: method meets problem, and how can the beholder avoid dealing with problems when the artist is convinced that his art is itself the solving of problems?

The acceptance of this metaphor by artist and critic has been hastened by the recent alliance between the arts and the universities. Formerly, scholars studied only the past and mainly the literary past. Their sole method was the historical and biographical. Now that the arts often live within and at the expense of college and university, contemporary works are studied, analyzed, and classified over and over again and in many ways. Some count images,* others pursue themes, still others discover myths in disguise. The art historians identify symbols and so do the critics of verse. The biographers apply to the artists and their works the psychological hypotheses of Freud, Jung, Reik, and other analysts of the mind. Each study confronts a difficulty which is set forth

* Attempts have been made to apply probability theory to the attribution of literary texts, and notably to the Dead Sea Scrolls. A distinguished statistician showed up the errors in assumption and in the understanding of probability which were involved, but this may not spare us the expense and agitation of a new scholarly "discipline." See Herbert E. Robbins, "Comments on a Paper by James Albertson," *Journal of Biblical Literature*, 1959, 347-50.

as if we could not live without its solution. Success in these researches may in fact lead to a degree or a promotion, certainly to a publication.

The "side-effects" are less happy. The "studies" of "problems" have often made what was dark darker and they have changed the use and worth of art. From a means of decent self-indulgence it has become a lion in the way: the world must now cope with art, square the mind with its innumerable messages and demands, even when no-meaning is the ostensible message. Scholarship cannot help making a document of whatever it touches, and documents are for scrutiny rather than enjoyment. The musicologist, taking mere "scientific accuracy" for granted as an old discipline, sees his work as that of a "pioneer, explorer, and discoverer" in the "research laboratory of music."[11] The art historian has his system for discovering the unknown, which consists in giving new interpretations of the "visible modules" of which paintings are composed; and when students find additional examples of the symbols thus traced, they talk of having "received experimental proof."

Not to be outdone, the literary critic protests that "the new way of reading . . . in prose fiction is much more complex than the ways of looking at plastic art," because the ordering of parts in prose is invisible: "We must take our bearings among the scrambled data."*[12] The difficulty explains why the methods of criticism do not bring agreement. Within any one type, such as the psychoanalytic, there are as many different results as there are practitioners. Among the critics who specialize in myth there are the literalists, who supply the means of direct substitution,† and the free associationists, such as the late Dr. Zimmer and his

---

* The new way of reading is, of course, retroactive, so that one can move into the library of classics armed with, let us say, "*Ethan Brand*, a Preparatory Investigation," by Kenneth Burke, and be set on the right track from the outset: "First, joycing the title, 'Heathen Fire.'" *Hopkins Review*, Winter 1952.

† E.g., Robert Graves, whose light and witty touch in his own poetry does not prepare us for this: "The myth [of Linus] has, however, been reduced to the familiar pattern of the child exposed for fear of a jealous grandfather and reared by shepherds; which suggests that the linen industry in Argolis died out, owing to the Dorian invasion or Egyptian underselling or both, and was replaced by a woollen industry." *The Greek Myths*, Penguin ed., II, 214.

editor and disciple Joseph Campbell, who entreat readers not to yield to the temptation of finding equivalents.[13] With them, green is the color of life and green is the color of death.

When art is conceived in these ways—as a palimpsest which yields a reading rather than a meaning—it must be faulty art or faulty method if the results do not "work out." But failure to work out in art or in criticism does not entail the same consequences as failure in science; there is no getting rid of what is shown up: we do not throw out Shakespeare even after his inadequacy is shown, and we cannot stop the methodic pens of divergent critics. Nor does anyone give up hoping for a definitive analysis; someday we shall *know* what *King Lear* means. We already suspect that the play is only "an expanded metaphor"— but which one?† No matter. The point is that criticism is no longer suggestive but declaratory; and when right, its relation to its object is that of science: it furnishes a has-shown.

Of course one must not press the analogy too far. The claims of "rigor" in "results" do not impose, even on the claimants, to the extent that they would have their readers accept. I mean that no one mistakes the contents of little magazines for science. Rather, the artistic and critical pretense works in parallel with the scientific influence. Nineteenth-century art brought a quasi-religious revelation; our age has turned the tidings cultural. Thus the acceptability not of feelings and hopes, but of modes of conduct, is decided now by reference to statements found in novels and poems. For the young aesthetes and their masters, the world of contemporary art has replaced the contemporary world—much in the way that for the physicist mathematical forms have replaced mechanical models.

How natural this type of translation has become was shown at the English trial of Lawrence's *Lady Chatterley* in 1960. All but one of the witnesses who might be termed academic or literary defended the book against obscenity by saying it was a parable of modern industrial England; the value of the work lay in what it

---

† Or can those students be right who say that it is Shakespeare's language that is doing the work of plotting and character-drawing, its own autonomous evolution having presumably brought it to the point where the author could retire and let his vocabulary boil the pot?

said about the machine age and the upper classes; sex was a symbol. So compelling was this interpretation that it was followed in the pages of the magazine *Encounter* by a polemic which applied the system in reverse. Dons and lawyers argued whether Lawrence was culpable for having left vague in that same novel a description of the love-making by which the hero "burned the shame out of the heroine." One group insisted on explicit sodomy; the others were satisfied that it was sufficiently indicated. The majority held that this failure to make the business clear showed Lawrence as a coward and a defective moralist and left the novel seriously flawed.[14] In short, art is a responsible agent to the last syllable, and its makers are held to a Draconian code of explicitness, consistency, and socially testable validity.

To be sure, there are moderates in both art and criticism who reject what current practice affords and assumes. But their words none the less show that modern works and the modern audience require exegesis, or at least "education."* And one regrets having to add that most often the critique or interpretation explains *ignotum per ignotius*—or rather, the un-understood by the unintelligible. Abstract and pseudo-technical language is *de rigueur*, whether in journalism or in scholarship, in the painter's catalogue or the composer's notes to his stoical listeners, in the defiant young reviewer or the essays of Pontifex himself.†

A defense, not indeed of pedantry and obscurity in criticism, but of studied inelegance for the sake of technical accuracy, was made a year or two ago apropos of Dr. Leavis' influential writings. His admirer listed typical words and phrases as the mark of the master: "discrimination . . . centrality . . . tactics . . . enforcement . . . concrete realisation . . . presentment . . . vitalizing . . . per-

---

* The debased form of the educational impulse can be seen in the notices on museum walls or in such program notes as that picked up by *The New Yorker* in 1962, which supplied for a performance of *La Bohème* the number of deaths from tuberculosis in Paris for the years 1901 to 1905, classified by age and social status.

† The journalistic: "The painting is freed of some of the formality which regulated it. It is lightly processed, warm in tone, etc. . . ." is preferable to the academic: "What was aesthetic alone became socio-aesthetic, a remarkable synthesis, not a symbiosis." And both are perhaps more excusable than the professional critic's indifference to prose: "We relate Hamlet's incapacity to a structural insufficiency, that is, to his failure to be part of an action-system."

formance . . . assent . . . pressure of intelligence," and added that the result was a "spiky, gray, abstract parlance," whose merit was to show that the critic "distrusts as spurious frivolity all that would embroider on the naked march of thought."[15] Whatever one's taste in prose, one is brought up short by the critic's own reason for "going abstract": the critic "as he matures in experience of the new thing . . . asks, explicitly and implicitly: 'where does this come? How does it stand in relation to . . . ? ' . . . And the organisation into which it settles as a constituent in becoming 'placed' is an organisation of similarly 'placed' things, things that have found their bearings with regard to one another. . . ."[16]

These words are a touchstone. They define criticism by affirming what I think is the fallacious scientific view of artistic complexity, which is bound to call forth all the "methods." No one disputes that some (not all) art is complex, but is it ever made up of pieces "placed" in an "organization"—which is to say a machine? To use my earlier distinction, is art not truly complex, instead of complicated as modern criticism assumes? If it has a true complexity, then it is "analyzable" only in a manner of speaking, for a limited and suggestive purpose, and not for ultimate certainty. We may speak of structure in art, but not of works of art as structures. To make a system of imaginary "parts" (rarely the same for two observers), to apply our little reagents and infer exact meanings, is to fool ourselves and the public. The motions impress only by resembling a technique and by giving results at variance with those of simply intelligent contemplation.*

What is also, in the end, bad about method and technique and structure and analysis is that they shroud art in affectation. The large public begins its aesthetic education modestly by learning historical and biographical facts, right or wrong, relevant or not. But the connoisseur goes in for refined explication according to a system. The ambitious newcomer to the arts naturally wishes to be graduated into this sophisticate group, and the two groups, in-

---

* Much of the nonsense about Method in the arts and in criticism arises also from the popular conception of scientific method as rigid and aimed at a single purpose. The philosophic unraveling of this confusion has been admirably done by Justus Buchler in *The Concept of Method*, New York, 1961, especially pp. 51, 85, 120, 149, and 163-64.

stead of using art for pleasure and wisdom, make it the excuse for pedantry and conceit. Wanting only the best, which they trust their method to discover, seeking purity of form or substance, they ignore the plain showing of art that it is mixed and imperfect, and hence discussable in ordinary terms.* Great works are metaphorically a world because they are large and thick, not because they are "autonomous" and irrelevant to *this* world; therefore to speak of them in anything but the simplest, clearest language is to be thinking of oneself and not of the masterpiece.

No gift of prophecy is needed to hazard the guess that these tendencies of criticism are reaching a dead end. A good many observers of modern art are evincing disquiet. Some social philosophers, much influenced by the methods of artistic analysis, are even issuing warnings in which it is comical to find the workings of method come full circle, as when a respected anthropologist deprecates the finding of symbols in too many places: "The tools, weapons, and instruments which man invents and uses for adjusting himself to his natural environment are things in themselves. Their form and function need no more meaning and no more interpretation by those who use them than knowing what it is they are and what it is they can and cannot do. They are facts in themselves. They do not necessarily stand for other things."[17]

After this it may be possible to read (and even to write) a novel in which the hero's name and lack of money are what they seem, being "facts in themselves" that "do not necessarily stand for other things." For it is clear that if each thing is not itself but a symbol of something else, then one can only speak of the original thing by finding a symbol for it in something different, and there comes a day when the substitutes are exhausted and one is forced to call a spade a spade.

To be sure, the worship of the artist which is now general, the

---

* The logical extreme of modern critical thought was the attempt, which made a stir in England some years ago, to provide a "musical analysis of music"; that is, composing around a known composition a musical commentary that should illuminate. It proved self-defeating, if only because the "relationships" discovered by this method would "relate" almost any themes and all particularity was lost, let alone graspability by the ear. It is characteristic that the inventor, Dr. Hans Keller, called his method Functional Analysis and referred to it as FA. See *The Listener* for Aug. 29, 1957, and the three months following.

importance of cultural products in trade and diplomacy, and the vast teaching apparatus that surrounds and engulfs art, make difficult any change other than the superficial one of style and fad. The Parisian innovators may sound a note of alarm as they ask themselves and the world whether there is not a "crisis of abstract art."[18] But the scientific culture will not on that account stop using art as an article of consumption, with no thought of its true place and purpose. And thereby the culture leaves the artist unpropped and perplexed.

This bewilderment includes the artists of high talent and conscience who have none the less yielded to the treasonable temptations and who now find themselves on the brink of pointlessness. As Yves Bonnefoy puts it for the most immediate of the arts, "painting, which in the past was a subjective questioning of appearances, has become a devout contemplation of depths at which all form disappears. . . . Today, the work of art aspires only to be a means—like prayer—for resuscitating and for reviving within ourselves a forgotten transcendence." After which it is a truism to say that because the transcendence is forgotten art is not strong enough to revive it.

On the side of transcendence, modern art bears an impossible burden, since it can serve but not replace religion, philosophy, and the springs of life. On its worldly side, our cynical or absent-minded employment of art forbids transcendent thoughts even if they should occur. Either way, there is too much art, it is too near for contemplation, and the machine—through advertising especially—provides a suffocating abundance of the "artistic," which cloys and weakens the hardiest appetite.

Nor does art withstand the extraction of its healing properties like Peruvian bark. The use of art for therapy is not new if we think of wise men's praise of books. But that was the therapy of pleasure unorganized. We now apply art like a biological.* Even outside the asylum and the hospital, it is part of the dietary. Every

* In certain theatrical groups, acting and directing are no longer much concerned with performing a play, but rather with working out the participants' intimate difficulties through it. The able director administers the therapy by addressing his remarks, under guise of stagecraft, to the several unconsciouses before him.

citizen is entitled to his share. A century ago Baudelaire was already noting the bad effect on the arts of popularizing their techniques.[19] He did not know when he was well off. What would he think of the "twenty-six primitive paintings by Harry Liebermann . . . who is eighty-four years old and has been painting for seven years"? This new artist, exhibited in a reputable New York gallery, "has had just one small showing, in Los Angeles, and at present is virtually unknown."[20] The touch of disapproving surprise at this neglect is suggestive of a collective madness that may find its place in history with the Children's Crusade, the Self-Flagellants, and the South Sea Bubble. Certainly, if one subtracts from the population of the West all those who live by art or starve for it, all those who study it or teach it or sell it, or criticize it or cure themselves with it, or enhance their status by rubbing against it, or express their national pride or private fantasies or intellectual aggressiveness by cultivating it, one is left with the impression that those who merely want it to enjoy would fit into a small hall.

But every folly has its reason, which it is never amiss to understand. The present madness about art is an abuse and a misuse by exacerbated souls that dimly perceive a right use. Dr. Johnson's maxim about literature, that its only end is to enable the reader "better to enjoy life or better to endure it," is the truth now perverted into a passionate identification with the artist. Not only his work, but also his life represents for those bound by techne the sensation of a crack in their prison wall.

To sum up rather than conclude: one must modify to the point of denial the idea of a rebellion by art and artists against science and techne. They detest and despise the scientific culture, they storm and rail at industrial civilization, but they are caught in the toils of a system that is as unitary as the idea of culture itself. By what I have ventured to call treason, artists share with science the faith in independent objects, the love of the abstract, the all-importance of method, the need for structure, and the fear of letting the subjective enter the game by way of humanity. Critics and connoisseurs are equally undone by the modern reality and seek again its horrors and their anxieties in their commerce with art.

On this common canvas are embroidered the flowers of evil as this evil is defined by resentment or projected by desire: violence, sex, the absurd, the primitive, the innocent and unconditioned. Since Nature continues to be artistically "corny" and excessively *déjà vu*, anything wild or perverse holds for us the presumption of rightness.* The fantastic in art resembles the steady improbability of modern science while also confirming the creative madness of great artists. Myths bear the same marks of high significance; so that throughout culture, whatever does not commend itself to common sense is on the face of it to be preferred. In searching at once for the primitive and for the improbable we are confident of taking the path away from civilization. And since we read man into civilization, we take all steps to abolish his identity, from making him face the distorting camera to showing him the derelict hero as model, not in compassion but in triumphant self-contempt.

---

* Compare Picasso's declaration to Augustus John: *Je déteste la nature*, quoted in John's *Chiaroscuro*, New York, 1952, 70.

XI

The Burden of Modernity

KNOWING the many moods and styles of art entitled to be called modern, I repeat that if the foregoing account harps on one note, it is because that one makes all the echoes now. There are doubtless other notes obscured today that will resound later and form a less obsessive tune. But what I have brought together should help to modify the common belief that art has latterly been isolated from society. An isolation ward that lets so much infection pass its barriers does not deserve the name. What is more, the passage has been in both directions: our culture is permeated as never before by the products and emanations of art, which act as continual reminders of our sorry state, permanent embodiments of the disharmony between our feelings and the terms on which we cling to life.

It is difficult, perhaps impossible, for an age to be clear about itself. If the western world were canvassed today for its opinion of science and techne, the great majority would undoubtedly express confidence and admiration. Yet it would still be true that those who admire science without qualification, who love the machine as a familiar, whose capacity for being uplifted by the swing and zoom of modern life is inexhaustible, have no effect on the temper of loud or quiet desperation. The voices of the hopeful are drowned out by the cries and curses of the despondent.

And among the silent, those who stop to search their hearts approve those who speak out to say that our towering civilization is in fact an abyss of degrading misery. It is not by accident or perversity that fame and prizes go to the novelists who show the absurdity of life; to the poets whose theme is the hollowness of men in an Age of Anxiety; to the dramatists who hold audiences by showing the pointlessness of existence; to the painters and sculptors who depict nothingness; to the musicians who avoid song and try to compose with the aid of chance. By using the name Existentialist, the philosophers assert that it is not from theory but from experience that they take *Angst* as an absolute starting point of metaphysics. The restless and perplexed try out the anti-western ways of Zen and yoga. Social workers and psychiatrists who cope with the now commonplace event of mental breakdown take for granted the "lack of meaning" in modern life; and these cries of pain are chronicled in journalism and conversation as "the crisis of the modern mind," or even more hopelessly as "the human condition."

Because it springs from the machine and its only-begotten state of soul, this conviction of misery must sooner or later affect all populist-industrial countries, although under Communism the emotion of collective effort for a time conceals the anguish.* The verbiage of Marxists interpreters cannot long impose on true minds, while in the West tradition is no longer a shelter to the intellectual. The French observer Jean Nocher, science-trained and in sympathy with the masses to whom he broadcasts, lists the intolerable hurts of modern life: it "hacks us to pieces by noise, by stupidity and folly on billboards, the screen, and the newspaper; by the anesthesia of jazz and the literature of nausea and nothingness; by pictures killing ideas; by the prefabricated intelligentsia and the misplaced indulgence toward others and oneself."[1]

What is difficult for the cultural critic is to discriminate between the age-old burden of life and the burden that is specifically modern; the *forms* of all our ills are bound to be modern. Again, though everybody groans, for whatever reason, it is obvious that people who read and reflect, and even more those who search out

---

* In a novel which made a stir some years ago, *The 25th Hour*, Virgil Gheorghiu damns both America and Russia because they have equally adopted the western notion of mass production for mass enjoyment, which "reduces man to zero." Hence the cold war is but "an internal technological disturbance" without intellectual meaning.

And Asia itself is not immune, if I interpret rightly a diplomatic traveler's remark that the way to make the Far Eastern peoples forget their mistrust of the United States is "to reveal the great secret that men suffer in New York as fully and painfully as they do in Rangoon." Joseph Lelyveld, "One to Speak, Another to Hear," *Columbia University Forum*, Summer 1962, 14.

and write, feel the pain more keenly and prolong it by addiction to art. It is also true that the city groans louder than the country and the seaboard than the interior, in keeping with the concentration of people, activities, and verbal intercourse.

But these variations do not obscure the central fact that began to be perceived a century ago: the connection of an indescribable malaise with the rise of techne and science. On this point it is appropriate to take as witness De Quincey, the opium eater who first anatomized the melancholy of the industrial city. Writing in 1856 about what it is that keeps a man from "dreaming magnificently," he ponders "the gathering agitation of our present English life. . . : steam in all its applications, light getting under harness as a slave for man, powers from heaven descending upon education and accelerations of the press, powers from hell (as it might seem, but these also celestial) coming round upon artillery and the forces of destruction—the eye of the calmest observer is troubled; the brain is haunted as if by some jealousy of ghostly beings moving amongst us; and it becomes too evident that, unless this colossal pace of advance can be retarded (a thing not to be expected), or, which is happily more probable, can be met by counterforces of corresponding magnitude, forces in the direction of religion or profound philosophy, that shall radiate centrifugally against this storm of life so perilously centripetal towards the vortex of the merely human, left to itself, the natural tendency of so chaotic a tumult must be evil; for some minds to lunacy, for others a reagency of fleshly torpor."[2]

De Quincey saw it all—the machine and the bomb, the abstractions ("jealousy of ghostly beings"), the pace and the magnitude, to which he adds in a following sentence the effect of numbers: Dissipation—the dissipation of thought and feeling from too many contacts with too many men. From that mid-century to ours the best men have been driven by "modernity" into the city of dreams, embracing tragic deaths through narcotics, literal or figurative, escaping to the South Seas or the asylum. But by the eighteen-nineties the evil was known to be not simply machines and "advanced" ideas, but the mechanical principle. Pater asserts that "all great poetry is a continual protest against the predominance of machinery";[3] as somewhat earlier Winwood Reade, the author of the remarkable *Martyrdom of Man*, had shown in a novel called *The Outcast* that a thinking man who reads Malthus and Darwin must wear mourning for humanity and ultimately lose his mind, as De Quincey had predicted.*

But it was not until after 1914 that western man discovered, in addition to the futility of his civilization, its depravity. The shock of a world war unexampled in destructive power, coupled with the physical and moral degradation of tens of millions—soldiers and civilians alike—opened the eyes of everyone to the paradox of progress. After the second world war twenty years later, judging minds could agree without concerted effort that any acceptance of "the system" and its pieties was a sign of either stupidity or hypocrisy. From this it was a short step to applying the *Non credo* to life itself. As Leslie Fiedler declares, altering slightly the sense of Melville's phrase: "All yea-sayers are liars."[4]

In earlier days, from Carlyle to Shaw, denunciation was argumentative and charged with intelligent hope. Now hope is absurd and the timely denunciation is repudiation, which does without argument or wit, as befits an age of rancor. A new quality, first developed in art, has thus entered the public domain of feeling, the mood of the delinquent. It is the blind resistance of the young who reject adult compromises and smash what they can, half confident, half afraid about the punishment to follow: Wilde's bored young men have regressed to Cocteau's *enfants terribles*.

In speaking earlier of men's love and hatred of the machine, I could not discuss the two feelings without giving the impression that there was a balance between them. But this balance subsists only if we take men and events by and large. The individual at war with the machine is often so oppressed and insulted and exasperated by it that no love is left.† For the machine is arrogant

---

* His other prediction, of torpor, expressed itself in life and letters as the pose of total detachment from morals, business, ideas, and even pleasure—the pose and creed of Wilde's people in *The Importance of Being Earnest*, which (though Dickens had given an inkling of it in *Our Mutual Friend*) rings the knell of Victorian industry and moralism.

† Arrested on a toll road for throwing nickels instead of quarters into the automatic coin box, a very decent citizen explained: "It wasn't the twenty cents—who cares about twenty cents? It was me against the machine age. I found a crack in the machine." The *New York Times* chose the remark as "the quotation of the day," July 25, 1961.

in its dumb unresponsiveness, worse than that of a brute. Even if a man loves his machine, as often happens, it is necessarily a one-sided attachment. *It* whistles and *he* comes running. When it fails, instead of a contrary will opposing his and accounting for the resistance, he finds only the caricature of an organism that he dare not even abuse. The possibility of kicking a donkey, slapping a child, cursing an enemy, will relieve a just anger, but I doubt whether anyone has ever seriously hit a machine, especially one bigger than himself.

We can really never accept the machine as rational. Its disorders have consequences of life and death out of proportion to the triviality of the cause: a gap between points or a speck of foreign matter stops dead a hundredweight of wheels and coils, and the absurdity does not seem laughable, as it should. Working or stopping, the machine can engender a helpless fury, and as a social product it draws to itself some of our hatred of other men. Perhaps the first machine to be hated by many as a personified evil was Dr. Guillotin's improvement for cutting off heads. He thought his "guillotine" would earn him the gratitude of mankind for preventing the executioner's inaccuracies, but he died disillusioned and execrated for a gift which had added a new mockery to death.

But our great and sardonic reward for introducing labor-saving machinery is that it has not lessened but increased servitude.* Looking at the bare results, it has lengthened life, but only by abolishing leisure. In setting up the pace and acceleration of which De Quincey spoke, it has made the supposed beneficiaries of the power and the device feel betrayed. There is no room, no variation, no escape; for machines have among themselves an affinity, a unanimity, that is not approached even by a one-party state. Machinery is totalitarian by nature; it maims and kills ruthlessly and impartially those that will not conform, needing neither words nor anger to utter its threats. Rather, it imposes its habits, like a drug; and

---

* John Stuart Mill noticed the failure of "labor saving" in the sixties, and so did Samuel Butler when he published the first sketch of *Erewhon* in a New Zealand newspaper under the title, "Darwin Among the Machines": "In the course of ages we shall find ourselves the inferior race. . . . Every machine of every sort should be destroyed by the well-wisher of his species." *Works*, Shrewsbury ed., I, 210, 211.

man is a slave before tasting so much as the illusion of pleasure.

One accordingly feels somewhat vindicated on learning that automatic toll collectors are fourteen per cent slower than human beings; or when one reads that to cut into the impenetrable regions of Brazil, man's hand must be used on the jungle before the bulldozers can function; or again, that the thoroughly mechanized trawlers of the northern seas are more vulnerable to the elements than the Portuguese dory fleet, in which single-handed fishermen off the Grand Banks have established a thirty-five-year record of complete safety.[5]

The coupled deprivations of work and of leisure are of course the most destructive and embittering.* For the lack of work means a lack of self-approval and of catharsis for aggression; and these lead (as I have shown) to the negation of any tolerable self and society. Lack of leisure, in its turn, squeezes and starves the emotions, for it takes leisure to feel, if feeling is to be charted and a right thought fitted to its fugitive contours. Nor does it bring comfort to know, as industrial man comes to know, that all his needs and troubles can be dealt with in the same way—by analysis and generalized assistance, the expert at one's elbow. Who does not grow depressed on reading "Scientist Urges Synthetic Food to Aid Hungry"? This latest notion is that mankind should batten on the inorganic. "Research," said the scientist, "could open an avenue for the chemical industry to relieve traditional agriculture production more and more effectively, and progressively replace it" by cheap synthetic proteins.[6] In the face of this utopia one does not want to be traditional, or even agricultural; one wants to feed the hungry with loaves and fishes, not keep them alive with slabs of chemicals —and this for the sound and selfish reason that deprived of their own food and having surrendered to the instinct for living at any price, they would no longer be the same kind of men.

When no such harsh necessity exists and, on the contrary, machine industry raises high the standard of living, that standard itself becomes a burden. Nothing today is more oppressive than the

* See the elaborate and discouraging survey made by the Twentieth Century Fund, *Of Time, Work and Leisure* (New York, 1962), of which the conclusion is that "the more time-saving machinery there is, the more pressed a person is for time" (p. 329).

compulsion to partake of all that mechanism provides under pain of being cut off from the experiences of the race. The standard of living is a host of material and social obligations. The advertisers press the point, but not they alone; the neighbors, the profession, the organization create a situation where in self-defense one must have a car, a telephone, a refrigerator, and the rest. Without them one's existence fails to mesh with that of others—shopping, illness, friendly and business dealings become enormously complicated; or else one must live as a quasi-hermit and be willing to condemn one's family to the same asceticism. "For the children" —that super-imperative—the standard is submitted to, and it yearly extends its scope.

The lack of ease that attends the simplest needs of life, like the lack of stability in the demands that techne makes on the mind, the body, and the pocketbook, is evident but not discouraged.* "It is trite," says an educational authority, "to point out that the world has changed tremendously during the past two decades, but the fact cannot be overemphasized."[7] Why can it not? Why, to consider the matter generally, is it not one of the worst burdens we bear that we find fact and word and thought and need repeating endlessly, mechanically, even the facts and words of change, of weekly and daily "revolutions" in things?

Strange that a diversified civilization, abundant in goods, rich in imaginative talents, should weary by an impression of sameness. Yet we receive it through every sense in the modern city, which tends ever closer to the uniform look, noise, and smell. The planner with his T square has large views of territory but skimpy ideas of variety. As an English architect put it, "Let us not be stampeded by the planners. . . . It is because every quarter of London . . . yet retains something of its past identity that this vast city is saved from being just a large waste of bricks and mortar."[8] And much of the sentiment recently shown by New Yorkers for such old horrors as Carnegie Hall and Pennsylvania Station is due to the

---

* According to students of the retarded, the increase in their numbers—highest in the United States—is not a natural occurrence, but a by-product of the increasing complication of life. Each new demand leaves some unable to cope with it who might have thriven modestly at an earlier time. See the report of the White House Conference on the subject, September 1963.

perception that they contribute to beauty by simply maintaining diversity of style.

The thrusting of sameness in our faces is the act of numbers. I have tried to show that the seeming ambiguity between large numbers—of people, of goods, of sensations—and the use of numbers to organize life is not a play on words but a natural sequence. Quantity blurs difference, rubs out quality, and forces us to distinguish by the mere sign of a number. The practice begins in trade and industry, where individual objects do not matter, and ends in the sort of statistical living which does not simply accept chance as the sufficient architect of our fates, but uses it as a substitute for the judgment of persons and their virtues. Thus it is an appealing modern scheme to assess individual merit and reward it in money on the basis of a simple computation of age and salary: the man who attains a high salary early in his career must have merit, and he thereby sets the merit index by which others in his organization are to be rewarded.*

The source of the pain here is not the likely injustice—all social systems fail to be just—the pain springs from the absence of a direct antagonist who committed or inspired the injustice. Man caught in the multitude feels he is at once a limb and a victim of the octopus. His struggles only tighten the bands around him. Democracy has given him equal chances, but the upshot is to make success a lottery, or else to rob achievement of its singularity, as when the equal right to education becomes the equal right to a diploma.

* The man whose salary is higher than that of anyone else of the same age has a merit index of 1.0; the man with the lowest salary has zero. It is then easy to tell the best without thinking. A gradual rise in merit index for those at the top indicates "merit by default," that is, the able have left, the rest were promoted. A rapid rise means "early recognition of high basic competence"; which, carefully read, is a tautology. But the statistical outlook soon loses sight of the ideas behind the figures. It can even remark, and forget the remark, that "the merit index is a valid indication of the individual's worth, or better, potential worth, only insofar as sound and flexible salary administration exists." At bottom is self-doubt: among so many we cannot spot merit—let us look at their pay checks and their gray hairs and combine the information: "The merit index is probably as good (or better) measure of over-all worth than is IQ a measure of basic intelligence." W. Shockley, "The Statistics of Quality Losses in Civil Service Laboratories," National Academy of Science, National Research Council, ARDC Special Study, October 1957.

Indeed, as equality of the purely numerical sort overtakes mankind, it affects all institutions, facts, persons, as well as all objects of interest. From being required to act and serve alike, they become interchangeable, until the spectator becomes unable to discriminate: all have the same features, all claim his attention equally, he scarcely knows whether he is listening to opera or having his teeth cleaned—equal obligations of a good citizen and perhaps simultaneous sensations of one sitting in a padded chair. The efforts of the advertiser to establish distinctions only thicken the fog, not only because the differences are often trivial or imaginary, but also because equality forces every person, group, and function to advertise itself if it is to survive. The most revolting effect of this competition is the publicity about bodies formerly outside commerce—museums, libraries, hospitals, and universities, and the services of the professions. The praise of these institutions is not always printed as advertising; it occurs in descriptive bulletins or is heard in public speeches at professional gatherings; it is self-praise, most often, the excuse of which is that even good things must "gain visibility" and sustain a "public image," that is to say, justify themselves by boastful appeals and promises; failing which, equality would blot them out and starve them.* And from this public confession of instability the individual can infer once again his own littleness.

The smaller man becomes in the multitude, the larger, obviously, the number by which his continued existence is expressed. It is not solely from the bank account symbolism and the absurd "Zip" code for delivering letters that the person is able to compute his fractional worth and mark his disappearance from the scene; it is from his increasing dependence on the knowledge of numbers for daily life. Automatic dialing, social security, the purchase of goods, institutional membership, licensed activities, dealings with government—all now require the numerology formerly reserved for

---

* "A short time ago, Mr. Pusey's personal assistant, speaking to an alumni group about Harvard's public relations, told them that 'President Pusey devotes a major part of his life to the question of Harvard's public image.' That's the way Harvard does it. It tries to satisfy the public that the job it's doing is worth while." L. L. Golden, "Public Relations," *Saturday Review*, Sept. 14, 1963, 76.

soldiers and convicts, and the reason is the same: when so many bodies have to be served, they have to be labeled and their wants reduced to system. With large numbers nothing can be carried in the memory, nothing can be left to chance, nothing improvised. To simplify would therefore mean to do without. With or without, the "I" in man is deeply deprived. "I have just received a government form," writes a despairing citizen to his paper; "my number is 6113406411121212072712. Could this be the real me?"[9] One may say, then, that the independent "behavior" of numerator and denominator makes us ill. "What is the matter with us, as D. H. Lawrence remarked, is primarily chagrin."[10]

But this is not curable through more experience and a juster sense of proportion. The torments of the time deprive us even of consistent desires. Literature and psychiatry dissect the modern loneliness, but we reject togetherness as a worse horror. Every complaint suggests that the speaker differs so much from his fellows that he cannot possibly be comforted by the few likenesses. The unique self, felt or wished for, is unacceptable to the ghostly surrogate of equality within us. We are lonely because we are *not* eccentric, willful, and egotistical. We lack the courage to be so; the unacted desire corrodes us within, and all we give out is the turbid anger that now permeates the written word at every grade of awareness.*

The thought of loneliness hides a wish for some clear authority, whether over our instincts, our waking dreams, or the thoughts of others. Colin Wilson, in telling of his unhappiness over the reception of his second book, and of the fury he has unleashed whenever his ideas have differed from those of other intellectuals (all denouncers of conformity), describes "the life I prefer." It consists in "listening to records for hours of every day, buying all the books I want, and seeing very few people."[11] This hardly seems the best posture for a leader of opinion, but the predicament is not uncommon. It was only a few years ago that Sartre wrote: "When one talks to people there are moments when one is

---

* "And as happened every time that he put his finger on his own mediocrity, he was overwhelmed by a sort of fierce pleasure." Françoise Sagan, *Dans un mois, dans un an* . . . , Paris, 1958, 175.

in communication with them. . . ; when one leaves, everything crumbles away. . . . It all adds up to saying I am alone . . . I'm not a leader—political or any other kind. I'm an intellectual who has taken a stand . . . not very propitious for establishing relationships with people."[12]

For intellectuals, one can see that the common burden of modernity is augmented by the sense of unfulfillable duty. How can they lead or reform a world which they disown? They disown it in part because its noise and numbers still any voice, in part because equality affords them no special platform. For a time the Marxist ideology gave the intellectuals of the West such a platform; that is why so many flocked to it. But the adventure ended in a disillusionment very like that caused by the bourgeois world. Marxism as a movement was corrupted by the same vices of scientism, jargon, make-believe, and anti-moralism, combined with heavy practical obligations. Among those who survived the ordeal not a few confess, like Avner, the author of *Memoirs of an Assassin*, that "after the ambrosia to which [reformers] are accustomed, all other food will seem insipid. . . . What are the symptoms? . . . a look of being drugged . . . a wry smile . . . a lack of interest in anything, and above all a lack of energy in dealing with the daily struggle of living."[13]

Even men who appear to have accomplished something arduous, often heroic, wind up feeling not so much weary or calloused as disgusted. Lawrence of Arabia is an example both as man of action and as writer, for he was among the first in high literature to revenge himself on his readers through the cruel depiction of the horrible and disgusting. Of himself he says: "I liked the things underneath me and took my pleasure downwards. . . . There seemed a certainty in degradation, a final safety. . . . The truth was I did not like the 'myself' I could see and hear." His sense of the burden is glimpsed in his remark: "Free will I've tried and rejected"; and the character he shaped for himself between disgust and determinism was, according to his biographer, "that of a man illegitimate and ashamed, ambitious and vain."[14]

Such a pitch of self-consciousness is of course unbearable, yet every emanation of science and every contrivance of mechanics increases it. They make a world where object and process rule: man an object to himself, a mere clutch of processes, is enslaved by the self-spectacle—to watch, diagnose, understand. The evil is not restricted to intellectuals; the simplest souls have learned the fundamental lesson of doubt: "You say you don't, but the psychologists say you do." The popularization of Freud has saddled us with more than clichés and constraints. It has engendered habits of analysis that form part of the burden—the habit, for example, of ascribing acts to motives secret and low, instead of to the mixed motives that sometimes become lofty by virtue of their spontaneity and intelligent results; the habit, again, of inferring too much from slight evidence—the love of the telltale fact which "debunks." The now-established belief that one's actions all come out of a common pot—the Id as defined by Freud, Jung's Collective Unconscious, the It of Groddeck—establishes also an equality of worthlessness in the acts and in the actor. This reductivism is congenial to the modern temper, and what it reduces is the individual.

By and large, the public does not know how to gauge the floating dogmas about the psyche—whether to honor the claim of each school that it speaks science, or whether to dismiss them all as vitiated by the counterclaims. What seems a firm conviction, because it has been preached for more than a century, is that man is an animal.* The arguments from bodily processes are there to prove this, and arguments to the contrary are set down as the anachronisms of religion or of the philosophic taste for transcendence. It seems at first liberating to be an animal, but soon it does nothing more than strengthen the grip of the standard of living— needs, needs, needs. It leaves no room, certainly, for that sense of mastery over body and impulse and even opinion which marks the philosophic mind.

Yet the proof of animality, which expresses a moral bias, rests on incomplete reasoning. A man who thinks might wish to be an

---

* E.g., the casual comment by the highly humanistic historian G. J. Renier, that the minds of certain higher mammals, such as dogs, "differ from ours quantitatively but not qualitatively." *The Purpose and Method of History*, Boston, 1950, 194n. Compare with this Samuel Alexander's skeptical view in "The Mind of a Dog," *Philosophical and Literary Pieces*, London, 1939, 97 ff.

animal, but knows he cannot. Not procreation alone but eating also differs visibly in man and the animals. Only by a crude abstraction of purpose and *process* are they "the same." As *activities* they are unlike. For where are the lion's drinking songs? Where is the goat's sonnet to his mistress? Where is conversation, legalism, fantasy, delicacy, deception, laughter, and vice in the conduct of animal feeding and coupling? The orthodox answer probably is that these and other human artifices are surface manifestations, irrelevant to the fundamental need; or still plainer, that they are illusions. One has a choice of explanations, but both alike make it hard to say what the study of society might be about: the "illusions" and "irrelevancies" *are* social and individual life.

The worst of the animal analogy is perhaps not the support it gives to the mechanical interpretation of human acts, for social life compels us to view them differently—*as if* we were responsible creatures like our ancestors. No, the worst effect is the insidious attack on private goals, which in a democracy are suspect to begin with. A man's ambition is absurd when men are animals, because as animal his only aim must be to prolong the race. "Nature," says the cliché, "cares nothing for the individual. It's the species that matters." The spectacle of a species capable of uttering this maxim and yet straining to go on with no other aim than going on does not raise one's esteem for those involved. Consciousness, one would suppose, breeds purposes as naturally as children, and if noble or great or delightful when achieved (like science), the purposes give meaning to the otherwise pointless succession of generations. But this logic science disallows, and no doubt to accept animality is for most men the path of least resistance. Only, it leaves them spiritless and unrejoicing in the daily round.

Because he finds little to his taste—all discontents and no civilization*—modern man likes to take himself down. But the creature within is cunning, and to pay back the self while proving its weakness, modern man devises many indulgences. One might instance compulsive overeating, alcoholism, and drug addiction, which are national problems in more than one country. But to speak more generally, if the will is dethroned, out of distrust of sovereigns,

* I borrow this apt allusion from my friend Quentin Anderson.

then each impulse may enjoy its minority rights. And if all men are subject to common drives, dubious in origin and result, then responsibility is diminished, as in children who do wrong before they can think, and must be forgiven so they can forgive themselves. Family life and schooling (and lately the law) have been conducted on these views until the general capacity for self-control has been noticeably lowered. True, the Second Man is well-rooted within us,* but we palter with him when we come to act: we are full of scruples but have no ethics. This is especially true in the intellectual world, where scientific investigators, teachers, medical men, psychiatrists, pollsters, purveyors of tests, surveys, and news, commit with internal impunity breaches of faith and acts of indelicacy which in another age would have brought private shame and social ostracism.

In the world of affairs it is this same laxity that gives color to the belief that we live in an age of immorality. By historical standards we do not qualify for the distinction, even though there is much bribery, cheating, and featherbedding.† Behind such misdeeds the attitude is the same as that which governs our manners —casual. An easy self-exemption from any code, a feeling that rules are unnecessary and rather absurd, that commitments are flexible and conventions arbitrary, serves to lift the burden from us for a moment. We extend this latitude to others when we labor to make learning, advancement, security, and honors easily available; when we protest against capital punishment and exempt from sentencing to prison seven out of ten of the criminals we convict;[15] and—less dramatically—when we see and use the fact that nearly every public regulation is not really meant: so many "crimes" are technical, they seem to involve no evil to particular persons—only inconvenience or affront to the mass. And if one shows goodwill, a brush with authority often resolves itself into the sudden surprise

---

* ". . . for together with, and, as it were, behind, so much pleasurable emotion, there is always that other strange second man in me, calm, critical, observant, unmoved, blasé, odious." Attributed to Lord Leighton.

† An expert on railroads writes: "In reading between the lines of some accident reports it is evident that normal responsibility to rules and authority has almost eroded away." Robert B. Shaw, *Down Brakes*, London, 1961, 459. It is equally evident in the reports of students cheating at examinations that the practice is no longer felt as a deliberate, danger-fraught assault on rules; it is, in the minds of the offenders and of their friends, a mere "eroding" of foolish authority.

of a true encounter between man and man. To get away from abstraction is a relief, which explains also the convivial tone of corruption in political and business life: the parties know themselves to be amiable and innocent—and everything else is formality.

No one can tell, nowadays, whether being conscious or unconscious is considered the better excuse, the better reason, for any act. And in this indecision, possibly, we touch on the worst of the torments inflicted by modernity. Every influence plays upon us to increase our awareness. Self-consciousness is the recommended state of mind; for we think it will save us from error. The stronger the mind, the deeper will introspection plunge to guard against the sin of not seeing itself as it really is. Irony is in readiness, to mortify from within and forestall contempt from without.* But having reached this point all prescriptions stop. No one seems to know how to make flesh and wisdom out of the augmented contents of consciousness. We stay put, spoiled for spontaneity and never tempered to finer uses, storing up fresh knowledge and suspicion about ourselves and the world without changing either. We are like learners at tennis or piano playing who should feel as awkwardly aware of their limbs after ten years as on the first day. So strong is the belief that this is the right way to feel that any artist of a former age who gives no sign of having talked to himself about his work is only grudgingly called an artist at all.† Though we are restive under the law, we do not consent to be children of grace.

From this ambiguity come still more discomforts, which belong to the realm of manners. There, too, we are self-indulgent and drop the constraints of punctuality, the minutiae of considerateness, the perfection of promised effort. We think it better for

---

* Thus a critic of autobiography: "The writers of confessional autobiographies are impelled to tell what they fancy they feel. And if what they fancy they feel is general enough so that other people recognize their own fancies and feelings in it, then these autobiographers are off the hook, so to speak." Virgilia Peterson, *Authors Guild Bulletin*, November 1961, 1.

† "One's main impression throughout is that Dickens is much more aware, careful, and delicate than has sometimes been supposed—at least in his treatment of detail. . . . [But this] cannot satisfy our doubts as to whether he was secretive, or inarticulate, or merely unconscious, about the basic idea and pattern of his fictions." P. A. W. Collins, in *Universities Quarterly*, November 1957, 108.

everyone to take his ease and then make up for it by the transparent arts of public relations—once again the mechanical device of abstracting a quality and serving it up processed. Not knowing, moreover, how to turn self-consciousness into second nature, we neglect the use of action and restraint in accustoming body and emotion to deliberate thought. We flee from Victorian moralism and would probably disbelieve Goethe's maxim that "there is not a single outward mark of courtesy that does not have a deep moral basis." We fail to see, on the one hand, that the Victorians' seriousness about themselves had something to do with their maintaining those selves under pressures akin to those now crushing ours; and on the other, that there is no use having sensibilities and subtle thoughts if they are not to make a difference somewhere—in the world or in our later self.*

If we turn from domestic to public manners, it is evident that industrial democracy has developed none fit for its use. Nothing but childish ways exist to soften the incessant collisions of people and their multiplied purposes. Yet we should take all means to encourage gentle feelings toward our fellowman by draping over his crudities. This has been done to some extent by putting attractive clothes within the reach of all, and so ordering charity that the beggar in rags and displayer of professional sores is no longer a common sight in the West. But this is not enough. The manners called casual reintroduce the causes of disgust which the better distribution of wealth tends to remove. Again, to oppose convention some make war against neatness and elegance, too reminiscent of the machine. Youth puts on slovenliness and age condones or adopts it, all forgetting the profound reality of appearances and believing by a kind of scientism that what is differs from what is experienced.†

---

* Compare a young Victorian writing to his fiancée from abroad to explain why he had not mentioned her in a travel diary sent to his family: "I consider my feelings toward you so sacred, that I should not like to parade them even to my nearest relations." Why not? asks modern youth—wouldn't it be *normal* and *healthy?* She's not an angel, but an *ordinary girl*, engaged to marry and *raise a family;* it's a well-understood *relationship*—who does he think he is?

† Shaw observed us early and well: "I doubt if there has been a country in the world's history where men were so ashamed of being decent, of being sober,

The burden peculiar to our times can be summed up by saying that a man no longer knows what to think of himself. This is familiar to modern psychiatrists as the problem of identity—the difficulty of understanding oneself, of being understood, of gaining the power to understand others. Anybody's simplest act or briefest mood seems enormously complicated. As indeed it is, thanks to the self-conscious analysis which is ever adding to the complication and preventing a return to natural complexity. But this problem of the person is linked with the general bewilderment of man at large. Because no graspable idea has replaced mechanical materialism, man continues to believe in the primacy of matter and suffers the emotional consequences: he believes that science has explained him by particles, by chemicals, by genes, by endocrine juices, by the unconscious, by economic and social forces, by culture. He is unable to sift or set bounds to these hypotheses, but their daily stress in spoken and written discourse molds his character: by default he regards himself as some sort of product.

It is this acquiescence that leaves us moderns helpless under the additional burden of education. Science imposes the burden of understanding; techne that of acquiring the results of perpetual inquiry. These results seldom give solutions; they more often define problems—tremendous ones. To solve them, what better than education? For as democrats we are not allowed to use force or relapse into indifference; we can only "educate," knowing the product Man is malleable. One could safely wager that everybody in the United States is at this moment educating somebody else about something, were it only about a trade name or a telephone number. Education in this sense does not call forth a positive act of will and mind; rather it is a form of conditioning, of product-making, which renders that effort no less indispensable and no less a burden.

Besides, curiosity is greedy—and suspicious: what do we not

---

of being well-spoken, of being educated, of being gentle, of being conscientious, as in America . . . although it is quite certain that the majority of them are doing this only on a false point of honor." And he adds that the description, which holds for all of America, holds for four fifths of modern civilization everywhere. *The Matter with Ireland*, New York, 1961, 67.

know that we should know in order to have our share?* And conversely, what knowledge could we pour into this organism to make it more useful to society? Hence the "retooling" of business executives; the capsule courses and charitable simplifications;† the imaginary new specialties; the theories and principles erected on top of theories and principles ("Education for Planning"); the bottomless restudying "in depth"; the career and marriage counseling; and the elaborate stratagems in school and society for educating to peace, happiness, and brotherly love.

The price we pay is the excess of words, schemes, and pictures. We are battered by demonstrations just as the Middle Ages were deafened by disputations. Not a piece of print or utterance on the air but shows by a string of propositions that something we never doubted or cared to know is as represented. This continual appeal to reason is an intolerable burden—on the eye, the ear, the mind— whether one takes it seriously or only strives to shut it out. Oh, for a truly hidden persuader! But there is none, because true persuasion, that which moves the sentiment of rationality, can only affect a self that is whole and for whom experience is vivid and immediate. The mind laboring under a load of ironies can only pick its way doubtfully among abstractions, which are well known for keeping life at a distance.

I have often impugned abstraction in these pages, letting the context explain what I meant by the term. It is obvious that the word means different things in the art of painting and in the science of behavior. And outside such particular applications, it is also obvious that no one can think or speak without using abstract words and ideas. Why, then, the repeated disapproval, to which is now added the judgment that abstraction is part of the burden

* "The local authorities of Poole, Bournemouth, [etc.] are being asked to help set up a mobile field unit to provide education and training for gypsies all over Britain. The appeal is made by Mr. Tom Jonell on the gypsies' behalf." London *Times*, May 22, 1963. When the gypsies ask for education, the end of the road has been reached.

† While pedagogues devise an alphabet made easier by omitting the letter Q, the Lawn Tennis Association is considering a change from the traditional words for keeping score because they are less simple than 1, 2, 3, 4; and the Anglican Church is discarding hymns (including Newman's "Lead, Kindly Light") because some of the words are ambiguous. *New York Times*, Sept. 20, 1963.

of modernity? It would seem there is no pleasing the critic of culture: he objects to factuality and to abstraction, to self-indulgence and to the stasis of doubt, to metaphor and to hollow propositions, to inquiry and to the lack of myth. The charge is true and is met by pointing out that it is the quality and fitness of the things enumerated that determine the critic's judgment.

Abstraction is indispensable but need not be ever-present. The modern use is obsessive, bound up as it is with science and business, with the democratic way and the statistical life, with the exhaustion of language and art and the peculiar make-believe of institutions. Abstraction in itself is nothing more than a pulling away of certain features of experience so as to remember and use them more readily. We abstract every time we ignore the particular thing to find ways of combining or comparing it with others. Each apple is unique, but by a mild abstraction we say "twenty apples." We are cautioned not to add apples and oranges, but we can say "forty pieces of fruit," and we can mix these with cars by the further abstraction of "objects"; and once again with ideas by speaking of "entities." At each stage, less of the first live perception remains present to the mind, and at the end we have left altogether the original apple that can be picked up and eaten.

There are, obviously, better and worse abstractions, abstractions that make clear and that make obscure, abstractions that are required by a purpose or context and others that are pretentious, redundant and evasive. Like metaphors, abstractions must be fashioned with skill and with sensitivity to their use and possible abuse.

It is by abstraction, as we saw, that "man is an animal": certain features in each group make the likeness plausible and possibly useful. But as the example shows, every classification threatens: it either omits the individual or forces him into a mold. And the risk is great that with familiarity the abstraction will be taken for the real thing.* This is what has happened in our times on a monumental scale. As the main stratagem of science, abstraction tempts

---

* Whitehead referred to this error as "the fallacy of misplaced concreteness"; it seems to me preferable to point more directly to the trouble by calling it "the fallacy of misplaced abstraction."

us daily first to conceal and then to forget what we are dealing with. Through the door of abstraction we enter a world of constructs, of "models," simulations of life, where we walk as in a dream, with heavy yet effortless steps. Objects are dim and fluid and distant, and mistakes about them may never be found out, but the air is oppressive and we struggle for breath. The dream turns to nightmare when we want at last to wake, to feel, to know, to do.

But the incubus on our chest can never be thrown off once for all. It is part of the burden, which it takes great energy to lift even a little. Consider: the abstraction is *like* the real thing; the statistical norm is *like* the individual case; the inductive law is *like* the descriptive enumeration. The urge to forget differences and neglect residues is very strong. The warrant for doing so is that science thrives by it. To treat the gravitational pull of the earth as concentrated at its center is an abstraction which is true, surprising, and wonderfully convenient. Again, it makes no difference that "natural law" no longer states what happens, but only what would happen if things were like the abstractions which science derives from them. Thus, the orbit of Mars is not in point of fact an ellipse; objects do not, in point of fact, fall at the same speed.* But in the lived life, as against science, the gap between concrete and abstract is like a pit we have digged for our undoing. Words betray us: when "mind" or "spirit"— abstract enough, heaven knows—become "libido" or "social trend,"[16] the already abstract thing turns into the abstract no-thing. Any connection with the person is submerged and lost.

By slipshod extension, abstraction becomes a lie. That is the guise it assumes in journalism and against which the mind must daily battle: "An Exchange of Videotape Ties World Together."[17] To get rid of such statements by concrete thought is a wearying task, but to give up the fight is to accept the reality of the nightmare: "A theoretical model is proposed for analysis of the nature of total institutions, an exploration of strategies of intervention

---

* It was often fortunate that early instruments did not permit accurate measurement, for the error encouraged intellectual leaps toward the fruitful abstraction. For example, Galileo's measurements of the pendulum did not disclose that in large oscillations the time varies with the amplitude of the swing. He was thus able to seize on the idea of equal periods in small oscillations.

(treatment) on institution, small group, and individual levels for achievement of therapeutic goals."[18] Such words as these are not aberrations of the vocabulary only; they define a view of life; whoever thinks them sees the world as a net of abstractions; he moves among the blobs of sense impressions like a subject taking a Rorschach test, looking for vague resemblances.

In such a world, which we all must live in more or less, the confusion is not due to too many ideas, but to their lack of solidity —De Quincey's ghostly presences. When, for example, someone abstracts from events to say "Teen Car Pranks Held 'Initiation'; Psychiatrists Liken Exploits to Zuñi Indian Rites,"[19] a part of the common world is spirited away out of our reach. We are not Zuñi Indians; our children are not reared to initiations; is their mischief mere high spirits or the germ of crime? That is what anyone concerned, anyone thinking, wants to know. But the abstraction, which is undoubtedly "a way" of considering the event, is not a way of picturing what happens in Connecticut where the study was made. Its very conclusion breeds disorder, since it shows that the speeding and daredevilry of youth cannot be condoned and the offenders wind up in jail or in therapy—which Zuñi rites do not entail. The whole "thought," therefore—and it is but one in a billion infesting our minds—is a gratuitous phantasm.*

It may be because we grope uncomfortably among these ghosts that we continually entreat one another to "face up to the facts" and "be realistic," and that we have borrowed and love to repeat the phrase "moment of truth," knowing that most of what we say has only the form of sense, and that it takes something like the bullfighter's vision of sudden death to arouse in us at last a genuine feeling and idea.

In present society one cannot refuse to have dealings with harmful abstractions. The machine abstracts from whole work; specialism abstracts from whole understanding. We have seen how the complications of business and the law abstract from the original

* How uncontrolled abstraction affects judgment may be seen in the many cases whose form follows that of a report on sterilization for birth control in China: "Only a few have failure in sexual functions, which however has nothing to do with the operation itself but is due to mental causes." New York Times, Aug. 11, 1963. Query: What is "the operation itself"?

concreteness of goods and trade. Planning—our great pastime and pride—is a series of abstractions projected into time and space, which frequently and luckily fail. Science, the maker of models, abstracts from the abstraction we call common sense. Language unfortunately follows suit instead of resisting.† Even its metaphors, as we use them, are abstract: they do not immediately yield their meaning but are images in roundabout. This feature is in fact a characteristic of our knowledge generally, which is not of but around its subject, like the circles in Ptolemy's system of the heavens. And in our perverse way, the wider the circle the truer and subtler we think the knowledge. Soon every situation will be simulated and dealt with at the second remove, for abstraction has biased us in favor of the inferential: we act not upon understanding but upon signals; and this, however successful, can never sustain the feeling man yearns for, the feeling that he is touching the real.

The burden of make-believe would not be suffered as it is if its contents formed a world of fantasy apart. It is in fact attached to real things—the goods and the satisfactions of what I have called futurism and utopia. And it turns out on inspection that these expected benefits, like the present standard of living, engender still other moral obligations. Their weight and their connection with "science" can be known from the frequent popular appeal of "heretical" books: "Is it unpatriotic to be lazy? Do you have to enlist in a strenuous physical fitness program? Do you turn down delicious dishes because you're afraid of—without really understanding—the 'cholesterol problem'? Is all smoking and drinking bad?" It is of course a physician who provides the "startling answers" under the rubric "It's Your Life—Enjoy it."[20]

But the doctor is wrong: it is not your life. It belongs to the community, whose commands contribute to a state of mind that might be named scientific fear. Protection, safety, insurance, prevision (planning), according to an ever-enlarging bible of ap-

† Nearly every public notice removes its subject to a distance: instead of saying "Do not talk to the driver when the bus is moving," it enjoins: "Conversation with driver is prohibited while vehicle is in motion." Compare in the same vein "CIA" or "USIS" with "East India Company."

prehensions, form the highest duty of industrial man. For one doctor who tosses off the cholesterol problem, there are a hundred who would conduct campaigns to make everyone aware of the symptoms of the chief diseases.* Already the citizen is assailed with pictures and slogans about crippling and killing ailments, but this is apparently not enough. And the point here is not whether lives can and should be saved. It is whether the fear and the duty inculcated by the propaganda are consistent with a tolerable life. When British Television broadcast medical and surgical programs, suicides followed. This is not complete evidence, but it is a sign that perhaps we have reached a peak of admonition beyond which it is not sane to go. No one has yet shown that to dwell on the thought of saving one's life is morally or physically healthful. The secular temper which condemns the former preoccupation with saving one's soul should study this and remember that the rules of religious salvation were at any rate simpler and fewer and surer than our medical commandments for the body.

In any case, the rewards of our worry are dubious. Longevity today is less and less attractive. Under the discordant name of geriatrics, we fumble with King Lear's predicament; and valuing the individual human life as we do, we cannot help rejoicing that medicine has prolonged the common span until centenarians have almost ceased to be unusual. But the plight of the old is a new spectacle of woe: they have life without a reason to live. They lack authority, respect, and responsibility. Unwanted by the business world, they are also unwanted by their younger families—one third of a nation superfluous for one third of its life. The conditions of rearing children, the irrational lack of space and leisure, no longer permit young and old to live together contentedly. This means loneliness matched by feelings of guilt: techne makes Gonerils and Regans as well as Lears. And when to this burden is added that of lengthened training for the young, the span of a career looks less and less inviting: the active man is forced to retire when his parents and his own children and grandchildren

---

* "The medicine [to cure half the cases of cancer] is communication. . . . The health field must find a way to get its message across." Dr. Leona Baumgartner, New York City Health Commissioner, in *New York Times*, March 15, 1961.

still need his support. What this prospect holds out to the love of independence and the zest for life is not much—a thin sandwich of stern activity between two husks of dependence and tutelage.

"Doing the right thing," in a scientific culture, is thus a duty that knows no limits. It is not moral alone but intellectual also, and practical—all-embracing, totalitarian. From the pain it causes, it inspires a fantastic tenderness to spare pain to others.* This excessive moralization of daily life is enforced by the anxiety about fact and the assumed resolvability of problems. But it must be added that it works on a peculiarly weak democratic self, one that does not lay claim to any social, religious, or intellectual tradition and discards it when born to it. The tradition of techne, if it can be said to have one, is that of the unhappy poor— the dispossessed and disheartened people to whom the world was harsh, and whose virtues are those of mutual aid—good nature, gregariousness, casual manners, mass tastes, and prosaic recreations. The moral ideal of such a culture is early Christian as opposed to Nietzschean: it is the ideal of *not* choosing between person and person, not denying anyone anything—of "privilege" made universal. This is why the moral pressure grows relentlessly as techne permits the diffusion of goods. It is not for selfish or sensual, but for moral reasons that ever more people feel obliged to be up-to-date, develop their personality, read new books (bestsellers are a "must"), assimilate the right food, music, movies, and detergents.

To discharge these duties is to pay a debt to the community, from which the self is scarcely to be distinguished: does one act upon those warnings of disease for one's own sake or to spare the community danger and the loss of its investment? Who shall say? Meanwhile, volunteering for PTA or hospital work, urban renewal, foreign programs, saving the children from toil and women prisoners from maltreatment, are but a few of the hundreds of legitimate demands made on the competent. One reads of oppressive feudal obligations, and one longs to be a carefree serf. No doubt it is worthy and important to "fight cancer" and cerebral

* A typical example is the demand that school marks shall not wound the feelings of those who fail. England has recently outdone this in the suggestion that schoolboy uniforms be abolished "to help alleviate the confusions of the teens." *New York Times*, Aug. 11, 1963.

palsy, and the rest in as many annual campaigns; but it is not un-reasonable to suppose that there is in people just so much moral energy; the fund can be overdrawn. And the combined weight of sermons in ads and problems in everything, of pursuing educational and other self-improvement (since to accept oneself unimproved is undemocratic), and of giving the community whatever it asks for, must sooner or later give rise to new forms of self-protection and self-indulgence, and further weaken the self.* For resistance today can only be evasive and hypocritical; we no longer acknowledge independence as a right, and do not care that mere compliance is no true gift.

In a corner once private but now almost a thoroughfare, the burden presses also, in the form of erotic cravings unsatisfied. Freud described in *Civilization and Its Discontents* the unavoidable restrictions that a high culture imposes on the instincts, and since his book other studies have attempted to expand or modify his cheerless conclusion.† The common form of the preoccupation is satisfied by pornography, vulgar or genteel, which a writer I have already quoted terms "the safety valve for people who feel the weight of modern life."[21]

Advertising, the movies, and literature of every kind join in this one "appeal," but the powerful effect aimed at is not matched by the results, which amount to a bored addiction. The death of Marilyn Monroe may have moved the western world because, as a grave New York newspaper said, she had been "standardized as a sex symbol";[22] but her own unhappiness leading to suicide would suggest that the emotion she incarnated was not exactly one of vicarious pagan fulfillment. It is doubtful whether instinct can ever be fobbed off with the distant satisfactions of the eye. The

---

* "Conurbanites" are not likely to be as strong-minded as the hoboes in an eastern city who were invited some years ago to assist a proposed study of Skid Row. Fifteen hundred volunteered. But on discovering that no doughnuts and coffee would accompany the interviews, the number dropped to a dozen or a score. Whereupon the investigators reported that if the group had so little social conscience as to fail to cooperate with a scientific project, they should be held in detention until they thought better of it. Fortunately, the law proved more just.

† The most notable is Herbert Marcuse's *Eros and Civilization*. Erich Fromm's *The Art of Loving* has enjoyed a wide circulation which is indicative of the common anguish, but it suffers from offering hope on too easy terms.

modern obsession with sex is rather the symptom of a lack and a search, and one obviously not limited to the simple sating of sexual desire. In fact, it is directed at the orgiastic, which is only in part sexual pleasure. Its other element is overthrow—tossing off the burden, not knowing or caring, heedless of consequence, thus paradoxically rediscovering the self by this strong purge of its ordinary cautions and controls.

Unless this indulgence is collective and ritualistic, it is sordid and it fails—as the civilized shamefacedly admit in their estimates of the periodic binge or the businessman's convention. These are not even the equivalent in catharsis of the medieval Feast of Fools, the Carnival, or other traditions that allowed the Church, the Guild, and the State to be mocked and flouted in willful confusion, wine, and ribaldry.* Our institutions seem the only ones in history that are sacrosanct without being sacred; so that in fact there is no safety valve; we keep ourselves perpetually under, drawing no breath but through a mask and a tube.

This, and the fact mentioned earlier that "science" has turned the sexual act into a self-conscious performance, graded, and "normalized,"† explains the step taken by Norman O. Brown in *Life Against Death*. Mr. Brown urges modern man to return to a pre-sexual erotic freedom, the freedom of the infant, rather than accept the resigned frustration of Freud or the emancipated sexuality of D. H. Lawrence. The aim is to rescue man from the prison he has made for himself during the centuries of science and machinery. The need is for "erotic exuberance," as free from the "tyranny of adult genitality" as from other compulsions, including the ideas of time and rigor, and the acceptance of self-control and aggression.

It is fitting that this remarkable book, which asks that we sus-

---

* From the sharp observations of Mrs. Trollope on the *Domestic Manners of the Americans* (1836), it would appear that the religious revivals and camp meetings of those days had an element of the riotous liberally tinged with sex. See her chap. 8 and 11.

† "Is Your Marriage Giving as Much Sexual Satisfaction as You Desire? *The Marriage Art* can help both men and women to improve their sexual performance. . . . It brings sex above the starvation level and gives it its proper place as a marriage *art*." Advertisement of *The Marriage Art* by Dr. John E. Eichenlaub, *New York Times*, Sept. 8, 1962.

pend common sense in favor of ideas easier to ponder than to act out, should be the work of a professor of classics. It disturbs the edifice of the scientific culture by digging below its foundations, and one may liken its central emotion to that which overcame the pagan world at the very beginning of Christianity. At that time also, it was the morality of the poor that triumphed over an Alexandrine sophistication, unshouldering the burden in the name of erotic exuberance. There are many signs, in our literature and our journalism, that a similar impulse to love simply, dumbly, clumsily, and without reproach is at work in many modern souls. The young express this yearning for *agape* in their dress and their intimate relations. The beatniks live by it. And we see the same tendency in the gatherings of Alcoholics Anonymous, of group therapy, group dynamics, psychodrama, and the like, some of whose practices are implicit in the work by which the Salvation Army met the misery of industrial London. But what Norman Brown desires is not a more or less contented industrial man; he wants our sick soul to be born again in the body of a Neapolitan lazzarone.*

Willing—indeed eager—as one may be to suspend common sense in order to follow an argument and see around the corner of convention, it is difficult to regard *Life Against Death* otherwise than as a further, and illuminating, symptom of our state. It suffers (as I see it) from the quite "scientific" fault of seeking the single cause and being consequently held to the single solution. It neglects circumstance, perhaps fearing triviality, which is also a modern fear; and it dismisses mind, as many have done since Proust and D. H. Lawrence. This rejection is inconsistent, for mind is taken at one moment as identical with science and the next as an organ wholly moved by unconscious forces. Even if the mind

* "Here tatters are not misery . . . filth is not misery, and a few fingers of macaroni can wind up the rattling machine for the day. . . . They are perhaps the only people on earth who do not pretend to virtue. . . . Yet what are these wretches? Why, men whose persons might stand as models to a sculptor; . . . whose ideas are cooped up in a circle . . . in which they are invincible. . . . Their humour, their fancy . . . their pantomime and grimace none can resist but a lazzarone himself. These gifts of nature are left to luxuriate unrepressed by education, by any notions of honesty, or habits of labour." James Forsyth, *An Excursion in Italy* (1805), *passim.*

at this moment is kept down in either of these ways, it cannot be left out of any proposal to lift the burden of modernity. For granting that the heaviest part of the weight may be lovelessness, the feeling owes half its strength to the way minds work in a scientific culture—the way of analysis, which forever divides and destroys. By study, by subtlety, by method, the breaking up of experience into pieces leaves each man with a tiny parcel of mental or moral real estate which he is continually bidding others to come and see and share with him. When they do not, because their own analysis yields a different geography, all conclude that there is no common world, that each lives and dies alone, surrounded by creatures more than half mad, incapable of love.

To this emotional insecurity is added the demoralizing menace of the two giants bred by techne—the Bomb and overpopulation. Mankind stampeded by too much life or too much death is the spectacle kept before the imagination by argument and rumor, and it does nothing to allay the desolation of existence. The wheel—if it was that inspiration which started it all—has come full circle: man is now as frightened of the universe as any savage we can name. Whether the terrors of hell in any age equaled these of our contrivance is unanswerable, except that those were balanced by the promise of heaven and the faith in a just God; whereas we have nothing so comprehensive.*

The cliché that man's misery comes from scientific "revelations" counter to his yearnings is absurd: it comes, on the contrary, from the lack of acceptable revelations, coupled with a prohibition against trying to fill the lack. And this absence of principles and objects to contemplate, of intellectual structures in scale with our facts, of tenable cosmologies, is in itself a threat of extinction: our own studies tell us that peoples die for lack of a myth to act out. And we have none. Even when the All was dark blind matter

---

* The just God was necessarily a personal God, who knew every living soul as a father knows his children, or better. This conception modern believers discard as unenlightened, and certainly as not in keeping with the generality of science or the probable concerns of divine energy. But without this fatherly particularity of God the reassurance he affords must be slight, and explanations by analogy with physics—"Man Termed Worthy of God's Attention: Those who feel they are lost in Cosmos should think of the electron"—carry little persuasion.

cooling, the story of the universe and of animal evolution had the outlines of a believable myth. But now that the door of the trap has been opened, and the ideas of energy and indeterminacy have brought back the chance of making a believable scheme for man to see himself in as both actor and spectator, we throw the chance away. We discuss factual minutiae, busy ourselves planning life underground, and go about persuading one another that man, "a needle in a star-stack," inhabits a minor object of no importance compared to the galaxies.[23] All this cannot help fostering the conviction that no conceivable life is worth living. The more heart and mind a person has, the readier he is to feel the force of these fears, commands, and privations, and to wish for nothing but an early end: there is clearly no use struggling to save a world in which one would find it hell to live when saved.

The instinct for bare survival is no doubt strong, but beyond this desire is the demand that life be held on certain terms: this is peculiarly human; it is what brought man out of the cave, what makes him die to resist slavery and oppression, what engenders all the talk of dignity which we know so little how to implement. For this reason alone, it would seem useful to ask how else the modern mind might look upon the scientific culture it has erected on the ruins of the theological, and how it could recast the role of science in life and in the mind.

No one, obviously, can single-handed supply such a reassessment or profess to chart a new direction. But even in a book exclusively descriptive and critical it may be right to point to elements already existing that could be used for such a purpose. They exist because they belong to a tradition still living, the same tradition to which we owe science, democracy, techne, and the philosophies of liberty and nature. Without these survivals contemporary life would be still more harrowing than it is. But their worth is obscured and their efficacy reduced, first by their variety, and second by their opposition to the passion of the age, which is to make science and the machine the only lawful expressions of mind. It may nevertheless be possible to show that this passion is not necessary, and that the opposition one might invoke and strengthen need not be an opposition to science and techne in their true being.

XII

One Mind in Many Modes

WHAT we discover as the cause of our pain when we examine the burden of modernity is the imposition of one intellectual purpose upon all experience. To analyze, abstract, and objectify is the carrying out of a single, imperialistic mode of thought.

That mode, the scientific, is fully justified—as everyone grants—by many of its results. It is a triumph of mind, a masterpiece of symbolism, an unrivaled satisfaction of the ancestral curiosity of man about his surroundings. Its activity is a magnificent spectacle, a virtuoso juggling with ideas, things, numbers, instruments. With superbly disciplined exuberance, its sober brotherhood works as no priesthood and no ruling class has ever done. And the work so richly and inexhaustibly rewards these seekers for their patience and genius that no man capable of understanding what science accomplishes can repudiate or try to dishonor it without giving up part of his manhood. So much is common ground.

But science has a hold on our imagination and an ostensible alliance with techne which make it today not a profession among others but the head and power of a scientific culture. The consequence, of which the main aspects have occupied us in these chapters, is a cultural contradiction: from the undoubted flourishing of man's science and invention springs the lament of men over their success: man is *not* flourishing. It may of course be said that among the lamenters there are few scientists and engineers; that is perhaps true; but the complaint stands, for the contented cannot be expected to deal with difficulties they have never felt. And though they are the ones through whom science grows and prospers, they do not own it. Science belongs to mankind. *We* made it

—we the merchants, craftsmen, theologians, scholars, metaphysicians, and manual laborers, we the poor paying taxes on our bread and toil. Science is ours to criticize like a member of the family.

As I said at the outset, and have just repeated, it is not the work of science in its purity that is open to objection, but the ideas and feelings and above all the habits which science generates and which, with our complicity, it encourages beyond endurance. Science in this responsible sense is answerable because it lives within our minds and is thus subject to our control. Not so techne, which has massive existence, like the mountains, and which as minister to men's physical wants no one can wish destroyed. Notre Dame la Machine may be without mercy, but our present enslavement, as we saw at the end of the chapter on our love and hatred of her, has a reason that man cannot disown and overcome: we must still hope to feed and clothe mankind.

Liberation in that quarter must come from other means than a changed philosophy. It requires practical resolves, a new discipline according to a new sense of propriety—in a word, new manners fitted to the needs of industrial man. The adaptation, if it comes, will be slow and difficult, for it will entail acting in the private domain in ways opposite to those called for everywhere else.

As a few random examples—for I am not redesigning Utopia—the machine must somehow be hushed, tamed, fenced off from the person at certain times and places. There are too many powered devices at one's elbow, too much that is childishly automatic, no matter how much we think we need conveniences to replace absent slaves. Again, language must be watched and the abstractions and bad metaphors (such as the cliché of educational talk, which recommends "full-time use of plant") must stay outside the ambit of the self, and particularly the home; the self being an end in itself and the home being a refuge from the world's work and a setting for intimacy and society.* Still more important, the minute economies by which techne, relying on the mass multiplier, makes millions out of mills must be barred from places where a manlike

---

* Arthur Balfour said a profound thing when he remarked that "Society, dead or alive, can have no charm without intimacy and no intimacy without an interest in trifles." Rectorial Address at St. Andrews, *op. cit.*, 325. Modern society has grown scornful of all trifles except the technical, and is without intimacy or charm.

existence is desired. No house was ever comfortable or beautiful that was not wastefully built. And we must treat with the same ruthless antagonism all planning that repeats and duplicates, all ersatz materials that cheat the sense of permanence, all sources of sound or image that destroy the privacy of eye and ear; all invitations to fit the body into shrunken spaces for any purpose, however transitory. The machine will be best served and best used when we constrain it within limits of our own designing.

Difficult to bring about or even to imagine as such acts and attitudes might be, they would afford at last that control of physical nature for man's good which has been so long talked of and so little experienced. But it is likely that the conceiving of new manners must wait upon that other necessary change, the reform of thought about science which would constitute the true scientific revolution. The link between the two is the attaining of a just view of man's place—not in nature, as we foolishly keep saying, but among his own creations. Though the new manners, whatever they turn out to be, should result directly from the Lesson of Statistical Life, they will, I suspect, not even be wanted until enough men have come to understand the Lesson of the Artist.

That lesson, as I tried to show, has been less clear and forceful than one had a reason to hope. The artists have gone sciency and perverse when they should have been steadfast and loyal in their opposition. But amid their confused words and attitudes one threat stands out: the overthrow of the intellect. Seeing how the scientific culture progressively drains the person of selfhood, to a point where the most intelligent give in and accept the behavioral idea that mind and purpose are illusions, the artists simply equate the scientific view of life with a diseased overgrowth of intellect and are ready to discard it as the cause and container of science and techne. From Proust through D. H. Lawrence to the present, Part I of this lesson is made plain.*

---

* Thus Robert Graves: "The pursuit of science . . . separates man more and more from his natural context in wild nature; and so far enthrones intellect as the greatest of all human attributes as to remove all checks on its irrepressible functioning." And he goes on to regret that Julian Huxley and Bertrand Russell "will not pause to consider dispassionately, in practical terms, whether myth and magic should be extinguished by science." *Food for Centaurs*, New York, 1960, 342.

But its repetition does not advance the cause, not only by reason of what I have called the artists' betrayal, but also and mainly because the proposed remedy—ten parts of science to one of art and creativity—is too abstract and simple-minded to cure the disease.† It is in fact a weak compromise: we artists will give up intellect to you scientists if you will allow us a little room for the imagination to consort in with poetry and myth. Or in other words, science is truth and poetry make-believe. Yet if the truth matters and its dwelling places are in question, the debate about it must continue. It is possible that science yields only one sort of truth. And debate, one would suppose, cannot go on without the aid of intellect.

Far from being a time for abdication by the vanquished, or for amiable give-and-take between the contestants, the situation of the western mind today is that of a violent drama in classic form: man forced to choose between his critical and his imaginative powers. This is the climax of the tragedy of science. As head and power, science is the hero whom we see caught in the toils of his own act, which may yet end in catastrophe for all. Science we have loved and revered, and like Oedipus it turns out to be our scourge. No need to show that when the mind is given up in despair, not alone by malcontents and eccentrics, but by the best that think and feel, science itself is in peril; its triumph is empty and the curtain will soon be rung down on the play—unless one makes a fresh start by asking what men were doing on this earth before analysis, abstraction, and objectivation put them in this self-destroying posture.

It is notoriously easier for a poet to write about the geography of hell than to show the way out of it. So we should not expect of the artists who depict our hell the mapping of our escape. Fortunately, we are not dependent on art or artists, nor on any doctrine or philosopher. There is in the western tradition a great body of thought and sentiment that antedates science and can never be ab-

† This prescription is roughly what Walter Gropius recently recommended on his 80th birthday. *New York Times*, Sept. 23, 1963. In an earlier article he had urged that "a compromise be reached between man and the machine," and that instead of "American quantity and diversity" there be "selectivity within quality." Interview in *Informations et Documents*, Paris, July 1-15, 1959, 34.

sorbed by it. The right name for those thoughts and sentiments is hard to give, precisely because our habit is to reserve for science and the products of science the reputable names. Thus if one says myth or poetry or philosophy or religion, one is tagged as a supernaturalist in whom the love of truth surrenders to easy comfort. A conflict is supposed to be going on between the stern discoverers of painful fact and the soft-fibered creatures who cling to the nearest fantasy. But this accusation depicting a storybook struggle is by now an anachronism and contrary to fact: nothing is easier nowadays than to believe in the conclusions and also the fantasies of science; nothing harder than to take a simple, unaffected view of the truths of poetry or religion. The predicament of the age is to regain the high ground where the thoughtful man can be at ease with his repressed intuitions and satisfy through many means his equally many capacities for reason and belief.

Before suggesting how this may be done, one objection must be removed. What if the present pains are only those of passing from a "prehistory" full of dreams and superstitions into a time of clear and tested knowledge? Would it not then be backward motion to rehabilitate the dreams? No friend of mankind can wish for a return to the fears and cruelties of the past, based as they were on provable error, even if today we suffer overmuch from fear and cruelty based on "truth." It is fortunate, once again, that the knowledge of our past renders the "growing pains" idea untenable. Such a passage from darkness to light might occur only if science could indeed master all that it attempts. But life, which spurs desire and fills the mind, is wider than science or art or philosophy or all together. Mind encloses science, not the other way around. Science is a little nugget within the mind, not a vast medium encompassing its thoughts. The mind is what we see with and cannot see around. We see its workings, but not how it works—in the sense, that is, of how it can work.* The capacity of a digital computer is known and

* C. H. Waddington, the biologist, concludes: "Biology cannot avoid discussing the nature of man, and one of the most obvious facts we know about man is that he is aware of himself. Since the nature of self-awareness completely resists our understanding, the natural philosophy of biology cannot but bring us face to face with the conclusion that, however much we may understand certain aspects of the world, the very fact of existence as we know it in our experience is essentially a mystery." *The Nature of Life*, London, 1961, 123.

can be told; no one can tell what the human mind can or cannot do.

This blessed blindness is inherent. There is no use saying, "Give us time; the explanation is incubating even now." Any explanation of the sort would be but another indirect report alleging "factors," "mechanisms," and the like which, enormously interesting and valuable as they are, do not touch the question they purport to answer.†

At this point a tendency of mysterious mind gets in the way of its honesty and truthfulness—the "passion for unity," which (as William James said) "is in some minds insatiate."[1] This passion has nourished the faith in science ultimately supreme, just as formerly it gave theology the same exalted place. The complexity of the human mind, its variability and headlong flights, obviously dissatisfy the seekers after oneness; they desire, as being more manageable and civilized, a fabricated unity, the singleness of an outlook and a method. No doubt the unitary desire has a noble side, as we know from the expressions of hope in Plato, Leonardo, Descartes, and others that through mathematics science would end disputes by showing the Truth about all things. That hope may be the scientist's necessary illusion to keep him at work. We have read moving testimony that this is so; the historical fact is that scientific unity does not last long: the scientific generations change and disagree, and (as Planck told us) not always by reasonable means.*

But it is not just because science is unfixed, tentative, often backtracking, and always unorganized that it is unfit for monarchical rule. Far more disqualifying is its irrelevance in a myriad situations that are of immediate moment to living beings. The facts that science collects, the models it invents, the relations it measures, are in those situations meaningless or disorienting. To live in society, which is to say in the relations of love, work, conversation, parenthood, or conviviality calls for judgments incom-

† The cliché that science tells How, not Why, is misleading. Science tells of proximate Hows. If, for example, "life" could be synthesized in a test tube tomorrow, we should know how to do it again. But *how* it is that the ingredients come to this pass, or how it is that viruses seem borderline objects between life and the inorganic, cannot be told, because there is always for the mind a leap between knowing the parts and knowing the whole: it is the same leap as that between nerve action and consciousness.

* See above, p. 15.

mensurable with those of scientific fact and truth. Can one imagine anything other than mass suicide if science and techne devised either a mind-reading machine or a true horoscope of accidents and deaths?

Moreover, in any sympathetic discussion of science one must at some point ask, Which science? Which science is to be accorded the right to impose its conceivings and formulations upon the rest of intellect? Not behavioral science, perhaps, which is as yet too incoherent. Between physics and biology there is, as we saw, a traditional misunderstanding: it would be hard to choose between them and not be misled. Within science as a whole there is the further rift between the materialism of the laboratory and the indeterminism of the theory required to explain laboratory results. And that, among other disparities, means for some that science does not produce permanent truth. The scientific quest does not discover reality, but is frequently rebuffed by it. This is a spur to finding, by fresh leaps of imagination, hypotheses that no experimental facts will deny, and then to organize these imaginings among themselves according to the rules of the game. Science at this point shows signs of abdicating its primacy and of climbing down to the lower status of, not art, but *an* art.

This is why, after the failure of unity, the next refuge for the observer who is no enemy to science is the dualism of the two cultures. We examined this notion in an early chapter: can we not, it asks, "balance" science with literature and the arts, each learning from the other, hands across the sea of ignorance? This is a form of the "compromise with the machine" advocated by Graves and Gropius which, besides overlooking all the specialisms that split science and art into hundreds of disunited groups, does not begin to satisfy the deepest needs of human beings. Art is not religion, nor has religion found unity. As for the love of nature, it is another of the innumerable manifestations of restless spirit seeking rest, for which a larger freedom is required.

Such freedom can be had, as I think, only on the premise that there is but one type of mind, the human mind, which works in a multitude of ways. These ways constitute not a dualism but a pluralism. Within it, of course, numberless distinctions forming temporary dualisms can be made; they exist for a momentary purpose.

One of these distinctions, first drawn by Pascal, can serve as the gateway to the realm of mind by showing the limitation which both science and all its opposites entail if adopted singly. Pascal observes that there are at least two ways of thinking—the geometrical and that which he calls the way of *finesse*. Being a mathematician and a writer of genius, he could speak from inward knowledge of both geometry and finesse. This in itself shows that the two ways do not belong to separate species of minds. Indeed, having regard to the ambiguity of the French word *esprit*, one should not take it to mean "a person with such a mind," but rather the spirit, the stance, of geometry—or of finesse—which anyone may employ alternately in dealing with the matters to which each is appropriate.*

Geometry (which stands here for science) takes up experience through rigidly defining entities that are stripped of the sensible aspects of reality. This is done in order to discover and demonstrate unchanging relations between these artificial objects and express the relations in symbolic form. A particular mind, unaccustomed to this sort of exercise, may find it hard to remember definitions that exclude so much, and may be unable to follow the chain of propositions. Such a person is distracted or repelled by the bareness of the ideas and consciously or not declines to regard experience as so nearly empty. But this failure is of one man's mind only: it implies nothing about "the mind" at large, the mind of man.

Finesse, on the contrary, is the power that seizes on relations by an immediate impression and without literalism. It takes up experience by a rapid synthetic glance, reaching truth, not through demonstration, but through an intuition of significance. This quick sense of how things go together and of what they mean may be right or wrong and usually it cannot give an account of its work: to prove its rightness or error is impossible; but this *finessing* of results remains the only means the mind has to know the world at first

* For stylistic reasons, Pascal alternates between *esprit de finesse* (the way or spirit) and *esprits fins* (people with such minds); but that he saw no radical division is shown throughout the discussion: "The reason that *certain* 'esprits fins' are not geometricians . . . and *some* geometricians are not *fins* . . ." *Pensées*, 1. In Pascal's day, it was not deemed impossible to be both poet and scientist.

hand. That knowledge is not *precise* in the way of geometric science but it is *exact*: one recognizes a friend's face without a chain of demonstration, just as one knows and responds to a song without analyzing its structure or statements.

In the realm of finesse the very things that science calls errors and that many worthy persons today are trying to root out of the common mind are sources of wisdom and strength. Our impressionable senses, our unstudied responses, our disregard of measurement (that enemy of "false impressions"), often save us from a stupid indifference, from the cynicism of the intellectualist, as well as from an excess or misplacing of feeling. For example, one person killed out of a million will seem a negligible event, though the impression will differ with the fame or merit of that person. But let him, though an unimportant stranger, be lying dead on our doorstep, and at once the negligible one-millionth we were dismissing shocks us into manifold awareness. Myth, poetry, religion, the mysteries and the Mothers do not suffice to exhaust the meaning of that encounter. Also true, the news that flood or earthquake has taken the lives of twenty thousand in distant parts will move us only indirectly. And by a further inconsistency of great social value, we shall think that a hundred starving people suffer twice as much as fifty, though to all eternity no more than one consciousness endures any given pain.

Reflecting on these ways of meeting experience, anyone can see that there are myriads of minds differing from one another incalculably, and not standardized types. That "the mind" is equally at home in science and myth, mathematics and poetry, music and the plastic arts, is shown by history and biography. If it were not so it would be impossible to teach children both nonsense verse and the multiplication table. Historically, it is the "same" mind that devised the alphabet and the game of chess; that built railroads and found the way to forge iron and temper steel, to ski, to fly, to swim by the Australian crawl; to teach the deaf-mutes to converse and the blind to read; that composed the Bhagavad-Gita and *Finnegans Wake*; that has accumulated gods and heroes, folklore, cooking recipes, medical knowledge, and the precedents of the common law; that solves crossword puzzles and the Japanese war code; that relishes the metaphysical poets and the clerihews of E. C. Bentley, that sees intellectual beauty in Chartres cathedral and architectonic splendor in Whitehead and Russell's *Principia Mathematica*.

It should seem unnecessary to repeat: the one mind of man does many things. In the haphazard list of acts and judgments just given, only a few employ the mind to produce science in the modern sense, that is, knowledge of the workings of physical nature, expressed in mathematical form, and based on definitions as remote from manlikeness as possible. What is astonishing is that the geometrical spirit can coexist with that of finesse within one culture as well as within one mind. For the spirit of finesse *assimilates* the world, grasping and re-forming it in shapes akin to its own, and it therefore ranges over a wider territory than the geometrical spirit, which begins by denying that kinship. If the geometrician makes a god of scientific method, it is a god in the non-image of man.

Yet there is an anthropos inside the scientist which no amount of Puritan discipline can keep under; and hence most of his life is ruled, like others', by anthropomorphic perceptions, hopes, and truths. He is and has to be a Pascalian double man without knowing it. This only heightens one's awe before the extraordinary will-to-believe that science requires of its men—to believe in its eloquent symbols, to believe in the concepts it has created, knowing they represent no independent beings; and finally to believe that in the course of time science, having dispossessed its rivals of authority, can remove them even from the mind. The lesson of Pascal shows us the one-sidedness of this understandable zeal. Geometry and finesse are realms of experience reclaimed from confusion by two different actions of the one mind of man.

But there is more to see in mind than Pascal's contrast of functions. The mind is not like its owner, a forked radish. Science and the manlike perceptions are not the only two ways, one rigid, the other flexible and more universal, of coming to grips with existence. The mind is more mysterious still. It is wavering and potent, obtuse and inspired, driven and controlled. A great jurist, Sir Paul Vinogradoff, once gave an excellent short description of man's conscious powers as they should appear to him: "A dog feels pain and pleasure, is moved to anger and joy, remembers blows and

caresses, may exercise cunning in achieving its ends, e.g., in opening a gate or in pursuing game. But its notions, desires and acts
spring directly from its emotions or from their association by memory. With man it is different. . . . The chords of our spirit are
touched from the outside by the various impressions made by the
objects we meet on our way, as well as by the physiological and
spiritual happenings of our own organism. . . . [And] reflection
makes it possible for us to rearrange our stores of impressions and
memories, to coordinate them in accordance with conscious aims
and deliberately selected standards. It is from this reflective element
that men draw their immense superiority over animals; that speech,
religion, art, science, morality, political and legal order arise."[2]*

This singular power, though it is not the whole story of mind,
is at once extraordinary and familiar. Yet how it performs, how
the possibility of its existence is to be conceived, *no one knows*. One
is at liberty to offer suppositions, but not to assert conclusions; and
the least satisfactory hypothesis is that of a mechanical contrivance.†
For that same mind responds or fails, gives up or recovers, as no
machine can be imagined to do. A psychiatrist who is critical of his
art, and who has lately questioned the ideas of mental health and
mental illness, points out that with the mind there is no predicting
with certainty: "Let us recall that insulin shock was also said to
have 'worked'; and psycho-surgery; and tranquilizers; and hypno-
sis; and psychoanalysis; and non-directive counseling; and existen-
tial therapy; and now group therapy. It is the tragic dilemma of
psychiatry as an applied science that in this field everything 'works.'
And nothing 'works.' But therein also lies the challenge."[3]

---

* Compare with this the doctrinaire hedging of a scientist: "Although man is
not the most highly evolved animal in all organic aspects, his psychological superiority cannot be disputed. This must not be considered an absolute superiority,
however, for most of man's mental advantages are also shared, although to a lesser
extent, by his mammalian relatives." T. C. Schneirla, Curator of Animal Behavior,
American Museum of Natural History, in *Encyclopedia Britannica*, 14th ed., 1960,
18, 705.

† To give one more instance of the general submissiveness to the idea of a
machine, which implies the desirability of tinkering, I quote a remark from a
well-known avant-garde poet, hostile to our culture, yet who apprised me of
"experiments" under way to use trance-producing mushrooms for "insights into
various disciplines—the first step forward in U.S. education in my lifetime, I
think." Private communication, March 22, 1961.

The challenge is not to the applied scientist alone. The philosopher or thinking man generally is forced to act or perish: the modern malaise deepening to disease cannot be cured until every man has made up his mind about his mind. What is true? How do I know it? Through which channels come the intimations I can trust? Answers to these questions are meaningless as long as a man takes, unexamined, the current clichés that do not so much settle the issue of mind for him as leave him casually wavering.* Far better when the issue seems to be settled by a scientist who angrily states in a polemic: "One thing I know: ideas don't move muscles."[4] For then we see in a flash how science cuts experience off at the point where it stops being manageable by the geometrical spirit. Break in on that same scientist and shout at him: "You're a liar!" and you will see the two fragments of experience reunited in the violent miracle of an idea moving muscles.

Side by side with the enigma of mind stands the question of what the mind can trust in its odyssey through Experience: where is truth and how do I know it? Freud, who offered his conclusions as science, has made everyone acknowledge the view first put forth in William James's *Psychology* that the human mind is naturally poetic and myth-making. Freud and his interpreters also established the idea of a mechanism by which the unconscious mind determines the conscious. Finally, this driven mind is understood to rule each person in the continual choice between the so-called Pleasure Principle and Reality Principle.

I am not of course summarizing Freud's psychology, but only referring to its cultural residue. The effect of so naming these contrasting "principles" was to reinforce the belief that pleasure is illusion and pain is reality—a notion in keeping with the Puritanism of science, which always suspects *anthropos* and his tastes. The mind in turn suspected itself, narrowly scanned all its intuitions, and

---

* When Arthur Koestler wrote a series of articles on extra-sensory perception in the *Observer*, the editor supplied a head-note in which he asked: "Is Man just an automaton reacting mechanically to stimuli . . . or has mind other dimensions too, which defy the Mechanist's approach?" This shifting of the burden of proof, like the absurd approach-dimension metaphor, indicates the stage of slothful indifference we have reached. *Observer Weekend Review*, April 23, 1961.

though knowing its native affinity with poetry and myth, concluded that it must not indulge itself in these natural pleasures. "Wishful thinking" became the worst intellectual reproach, as if "reality" were not safeguarded by curiosity, which is another of the human mind's powerful appetites. The very caution inspired by curiosity grew into an absolute commandment: Thou shalt not pay heed to anything but the real, and the real is the painful.* Since it would be also painful, if it were practicable, to discard all untested memories, hopes, and beliefs, the modern psyche is tossed about amid conflicting temptations: "Science" tells it to recognize the "demands of the organism," no matter how gross, selfish, or even immoral: to deny them is held to be like refusing to lubricate a machine. At the same time, the "reality principle" compels it to deny the simple, selfish, and possibly immoral demands of the mind.

This inconsistency is part of the reason why D. H. Lawrence thought that "we can never recover our moral footing until we can in some way determine the true nature of the unconscious."[5] But if the unconscious is what we think it is, it has no "true nature" —it is capable of anything and our accommodation cannot be *to* it so much as *with* it to whatever we deem right and real. Carlyle was nearer the truth when he said that "our spiritual maladies are but of Opinion; we are but fettered by chains of our own forging, and which ourselves also can rend asunder. . . . [The evil] comes from our unwise mode of viewing Nature."[6] At present, the "immoral" demands of the mind lead a hole-and-corner existence. From astrology to psychical research, from nature-loving to asceticism, from the scholarship of myth to the public respect for poetry and art, from conscientious drug addiction to the old and new religions of East and West, there are only minority refuges open to the inhabitants of the scientific culture.

These activities have no place in the prevailing scheme of the real; they exist on sufferance as eccentricities, harmless anachronisms, or anodynes.† They are not considered modes of knowing, though

---

* Our poets take up the chant in counterpoint: "Human kind," says T. S. Eliot, "cannot bear very much reality." *Four Quartets: Burnt Norton.*

† Still, there are degrees among them: recently the astrologers of New York City petitioned for the suppression of less learned soothsayers. The Commissioner of Markets, as the guardian of advertising, declared that he would investigate "these phonies claiming to be astrologers." *New York Times,* Aug. 8, 1962.

they often form the subject of scientific study.* The practice of translating and interpreting experience by reducing it to something else not directly present to the mind disposes of all these modes of thought, ostensibly forever.

Yet the great fact remains that the mind continues to "take" reality in myriad ways, ways which, after all errors have been sifted out, cannot be graded or measured against one another. The newly rich in scientific lore insists that water is H_2O and cannot be those shining diamonds that you see on the lawn. He is sure that the birds are not really singing, hopping, and flying about in a delightful way, because all they are doing is mating, feeding, and avoiding their enemies. His monogamous attachment to one idea is at the cost of putting his senses and his candor in escrow. For the sake of pursuing one sublime entertainment of the mind, he prohibits other, no less admirable entertainments and drives us to exclaiming with Blake:

> May God keep us
> From single vision and Newton's sleep.

The single vision, no matter which, will not satisfy all demands. Time passes differently in jail and in lovers' meetings or dreamless sleep, and the clock time of a play bears no relation to that of the action seen. The law, again, is forced to "take" in its way, and not in science's, the time relation of two events: it decides otherwise than Einstein would the question of simultaneous deaths, for science sees them as processes of possibly unequal length and rejects simultaneity that is not of one place.†

On many points, again, the message of science unsettles common sense and gives only vague auguries in its place.‡ A man who drops a cube of sugar into his teacup has the right to consider it a simple and intelligible act; the scientist has the right to tell him that the dynamics of that dissolving are extremely complicated, virtually

---

* See above p. 190 the mention of Dr. Dement's experiments showing that human beings need a nightly quantum of dreaming, failing which they will make up the requisite amount on succeeding nights.

† See the Grosvenor succession case, which went to the House of Lords and on which the Lord Chancellor made some pertinent remarks about the inapplicability of science. James Avery Joyce, *Justice at Work*, London, 1955, 124-27.

‡ As an astrophysicist once said disarmingly in my presence: "It is almost easier to explain certain phenomena than to accept the explanation."

beyond anyone's grasp. We have no choice but to trust whatever knowledge seems to fit. It may be Newton's laws or some deliverance of the past century or the previous ten thousand years: we need it even if the mind has been truly touched by the finer thoughts risen from contemplating the atom and the galaxies. For moral as well as practical reasons, a man must remember Huxley's fortitude and not be afraid to see the sun *set*, regardless of what he knows about the rotation of the earth. Nor must he let himself be hourly shaken by the restless horde of transferers of scientific ways to the affairs of daily life. The farseeing Bacon warned against what has in truth happened: "I foresee that if ever men are roused by my admonitions to betake themselves seriously to experiment . . . then indeed through the premature hurry of the understanding to leap or fly to universals and principles of things, great danger may be apprehended from philosophies of this kind; against which evil we ought even now to prepare."[7]

But (it will be said) what about foolish beliefs and ancient unlikelihoods—spiritism, telepathy, faith healing, visions and miracles, which have more often harmed mankind through fanaticism than they have sustained it through flattering hopes? How can we give them the right to coexist with scientific truth? Think of the swarming dangers if cranks and the credulous are given the slightest encouragement, even indirectly. In short, beware: the specter of Subjectivism also haunts our culture.

The full answer is twofold, but not all thinkers espouse both halves. Many simply challenge the imputation of foolishness and impossibility in these beliefs. Telepathy, they say, is recorded fact. Millions believe in the religious miracles and visions by which the saint is often signalized. Faith healing occurs not only in so-called primitive societies but at Lourdes and in Protestant churches of midtown New York. A Canadian physician, himself no supernaturalist, acknowledges the success of a Nigerian witch doctor he observed.* Physiologists fail to understand how the Papuans survive on an almost exclusive diet of sweet potatoes. Anthropologists testify that when certain ideas depart the people perish—and this in no metaphorical sense.

* "Healing at Altar Told by Pastors," *New York Times*, Sept. 23, 1963; *Toronto Standard*, March 24, 1963.

Thus, the first argument against the primacy of the provable is that the unprovable and even the doubtful are necessary to man. What is correct and demonstrable is good to have but it is not enough. And sometimes it may be too much because it leaves too little room for its contraries, imperiling life. The poets have said this in countless ways, and shown it in their lives by naming the illusions that helped them to survive or create. Like the statesmen they have made the obvious point that without prejudices, partialities, and arbitrary attachments there can be no society, no family, no friendship. Mother love will forever elude laboratory tests and can rarely prove its case; yet we know it exists and that it is a fine kind of extra-sensory perception. It belongs to that class of feelings that Wordsworth, speaking of our attachment to objects that have long pleased us, calls an "honorable bigotry."[8]

So, to the cry of Subjectivism, Superstition, Unreason, Obscurantism, which the poets' view always provokes, the poets and their friends reply that, thanks precisely to science, modern man can now choose between geometry and finesse with a superior lucidity about each. He has had a strong dose of geometry and he knows that the attempt to think exclusively by it has turned man's sensibility inside out, making callous what should remain tender, and spoiling life. He asks himself which is better, the rule of method and number that will bring men to "Newton's sleep" or the compromise that is hospitable to error and folly for the sake of a waking life?

This dilemma suggests the ground for the second half of the answer to those who fear the subjective. That ground and that answer are characteristically the historian's, for the historian understands that there is no such thing as unanimity, even where it appears to prevail. To his disciplined eye science is no more at one with itself than any creed, despite the excellence of the scientific tongue. His skepticism saves him from the belief that mankind is moving toward the hoped-for unity in which disputes will cease and the truth be held fast.* He knows that mind exists in many modes, which are communicable and yet which form "families of minds" that seem disparate. Being sensitive to the particular coloring of

---

* As the master skeptic said: "There never were two opinions alike in all the world, no more than two hairs or two grains: the most universal quality is diversity." Montaigne, *Essays*, II, 37, last sentence.

truth and the uniqueness of men and their purposes, the historian is bound to champion the variability of truth. He respects it with a piety which does not exclude but limits inquisitiveness. He is not taken in by the generations' recurrent "now we know." He remembers witchcraft and phlogiston, vital force and the luminiferous ether, Piltdown Man and therapeutic bleeding, and he wishes there were less *hubris* among the minions of intellectual authority.

But few of us today draw comfort from the historical fact of history. We engage in a futile search for consensus on great matters, while we recoup ourselves by pointless diversity in small practical ones. We labor to make true the hypothesis that, except for aberrations, life appears the same to the three billion people on the earth's surface. And we add that if it does not, it ought to, by way of ushering in the utopia of the adjusted. Meanwhile there is fear that unless all minds feel the same difficulties and respond to them alike the world will fly apart. This notion of response ("reaction" is the current word) hides an analogy with the machine, whose parts fit and function together in only one way. But how many in each generation are likely to venture very far from the serried ranks of the machine-pressed minds? Is it not because commonplace thought is common and guarantees stability, while original thought is rare and disturbing, that a democratic society preaches the virtues of individualism and also punishes the sin of individual thought?

Still, the self, even the most moderate, has many interests and requires more than latitude in knowing and taking experience. The illusion which I have said it needs must, like other desires, be judged by results. Men of integrity have cultivated illusion for purposes their reason approved: to cheat pain, for example, or boredom; to avoid failure and despair; or simply to achieve the unattempted. If superstition can sustain creative power, as the poets say, the need of fiction may be a generality for conscious life. The modern world, certainly, seems to need the unconscious and the electron—equally suppositious—in order to live in the way it has chosen. I have not mentioned dancing or daydreaming, which are also activities of the human mind that enable us to know what is, including ourselves. For one is told that if one acknowledges such facts aloud, one will "open a very serious door" to the horror of private universes, to hallucinations—in a word, to heresy.

True, historical reality—to use Margenau's doubly apt designation—does harbor innumerable errors and fancies. Many are untraceable or never to be caught up with. Others come from science, which in its diffusion scatters them faster than they can be removed. Still others are policed, that is, they become institutionalized illusion, as in the usages of advertising, the operations of credit, and indeed all the amenities and decencies, which rest on a reciprocity of fictions. Yet none of this has ever kept men everywhere from wanting just as strongly to believe only what was so: the Truth. They have found it hard to fix; they have been dismayed by the difficulty of reconciling the endless "takings" of experience of which the mind is capable. This pluralism of truths is the lesson of history. Part of its teaching is that in the diversity of "takings" few are obviously arbitrary and culpable. What in our disputes we keep calling a half-truth is usually a whole truth, but one to which we can add a second and a third. To heed them all, each in its place, would presuppose a godlike capacity. But we can strive in that direction; as Shaw said when he was accused at a public meeting of speaking like two different men: "Why only two?" A universal man should approach within himself the universality of all men. Though he cannot take in all possible conclusions, his mind should be open, at least temporarily, to all modes of use, from the analysis of the scientist to the synthesis of the historian, from the art of the poet or musician to the study of primitive man's superstitions, and thence upward to the intuitions of mystics and religious men.

To put the case historically, the twentieth century needs that multiplicity of consciousness which Henry Adams chronicled and for which he mistook the agitation of post-Darwinian thought. This multiple consciousness already exists within science itself. We know that physics and biology overlap but also diverge completely as they grip their unique subjects more firmly. Within physics, the reversible formulas of wave mechanics, which relate to particles, propose a different world from the irreversible formulas of thermodynamics, which relate to the particles organized into molecules. In chemistry, the models vary deliberately and inconsistently with various analytic purposes. In shifting its tactics, science follows nature in more than the ways of expediency; it follows the fickleness of nature, which is anything but single-minded: sharks are fish, not mammals, but many species give birth to live young. This doubtless pleases the Great Architect, though a planner modern style would frown at the irregularity, his mind being (as we say) conditioned to system.

Nietzsche, who witnessed the jubilee of science in overdisciplined Germany, saw the need of an intellectual overturn. He had the courage to question the traditional etiquette about truth and method and fact. He asked why "the certain is worth more than the uncertain." He smiled at "the consciousness of triumph in every inference." He thought naïve "the belief, implied in concepts, that things are intelligible."[9] He called for a new type of thinker, who would unite in himself the previous special types—critic and historian and riddle-reader and moralist and seer and dogmatist and liberal—a thinker who would honor truth yet stay free of its clinging embrace: a Montaigne or a Socrates.

By the extent of its subject matter the historical outlook necessarily surrounds and "places" the scientific. Here too it remembers: it reads in the record how each period has had the kind of science it wanted—the mind of the seventeenth century molded to a mathematical form, the eighteenth to a mechanical, the nineteenth to a biological. By a shuttling of cultural cause and effect, the abstractions proposed by men of science seemed to follow the prevailing style. When matter reigned supreme in the nineteenth century and waves were wanted to satisfy the suppositions about light, the luminiferous ether "was supplied to perform the duties of an undulatory material."[10] Earlier, when the belief in God was still vivid and the term "spirit" was therefore intelligible, Newton supplied the want of an ether by calling Spirit an entity which served at once the needs of gravitation, cohesion, electromagnetism, and the bodily exercise of the will, all conceived as so many vibrations of this medium. For modern physicists, the characteristics of the space-time continuum help out in kindred ways, despite scientists' disagreement about the definitions of space, time, event, creation, and being.

The student of history moreover knows that the present conflict between man's claim to independence and his urge to call himself a machine is the third time around in the same whirligig: the western mind has been put through the ordeal once in each of the past three centuries. After the first round of the Enlightenment, it was the upheaval of the masses in war and revolution that saved the sophisticated by showing what action was. Even so, the younger (Romantic) generation at the turn of the century had to go through painful crises to overcome the dogma which taught them that they and nature were neither sentient nor alive. By the middle of the nineteenth century the compressor was again at work and now, a hundred years later, we are in no better case than the first victims, the Englishmen who petrified the teachings of Locke and Newton and whom Bishop Berkeley helped to reawaken.

Today, the liquidation of Romanticism, necessary on other grounds, has left us unwarned and directionless. All we have as legacy is a pitiable sexual emancipation. We lack the men of intellect who between the 1880s and the first world war began to deal with this obsessive question of mechanism. This is to say that, as in the arts, the proper work of our age has been hampered by the great cultural break of 1914-1920. We nod to the names and doctrines of the time of transition, but we do not feel their force, we do not give them the assent or dissent which would make a difference to our thought: the whole pragmatist movement, from Nietzsche and Samuel Butler to William James and Bergson, is misknown.* The bearing of the works of Boutroux, Poincaré, Meyerson, Duhem, is felt as tangential, "interesting," but not pressing on our minds to reject or accept. And of course no works of literature, such as André Beaunier's novel-parable of 1912, *L'Homme qui a perdu son Moi*, have survived to goad us into thought on its theme: "Science, admirable science, cannot alone manage the life of persons and societies."

Each man has a moral duty to find out what is true, but he is bound also to keep hands off his findings and ponder them gravely if they do not accord with previous ones, his own or the pundits'. And to exercise this self-control it would be well if instead of a single dividing act—"I believe; I disbelieve"—the mind were

---

* For example, no sufficient distinction is made between James and Dewey, who in some respects stand to each other as finesse stands to geometry.

trained to recognize degrees and qualities of belief. The model here again is given us by the poets: Goethe, an excellent scientific mind, did not believe in witches to the same extent that he believed in the metamorphosis of plants. Wordsworth trusted his sister and impulses from a vernal wood differently. We acknowledge that a musical work does not always "convince" in the same degree. The quality of belief in one's own existence differs from the several qualities of belief that characterize for us the existence of friends, of acquaintances, of public figures.

If, at any rate, there is to be a redirecting of the mind, propositions will not suffice: the whole gamut of belief must be brought into play. For, as Cassirer reminds us in speaking of symbols, meaning attaches only to what "appears important for our wishing and willing, our hope and anxiety for acting and doing."[11] Now Techne is not going to give back to us the willing and doing once linked with Work. The one activity we can recapture is of the mind, and this presupposes a belief in its powers, in man whose mind it is, and in the multiple vision which is the proper exercise of mind.

In the call to this renewed action the words used must connect with the motions of the soul that Cassirer names. To be sure, we already hear many urgings and many instructions as to what we must do if life is once more to be worth living, if science is to be our servant and not our master, and if we are to avert the menace to every man that man now is, when he can reshape his progeny or destroy his race. "Can we find in ethics," asks the typical inquirer, "the humility that will make that power safe?"[12] There are rhetorical questions that make one want to turn on one's heel, and this is one of them. Other questions, put by critics of scientific culture —religious thinkers, students of parapsychology, earnest novelists and bewildered painters, and even our specialists in social affairs— come closer to what is wanted. But too often they take their tone from what they criticize and address us in its dead language: "The sociology of knowledge, by employing a wide variety of sociological systems, has shown that alternative explanations of the same phenomenon are possible, indeed can facilitate a complete understanding of any given problem."[13]

This contains the idea of the multiple and simultaneous vision which I am not alone in thinking we need; but those are not words to rouse the torpid. If any proof were needed that science has usurped the place of religion, philosophy, and myth, the absence of audible voices would show it. Our babel is a mumble. Straining an ear, one perceives that thinking men have let the speculative intellect shrivel up within them. Because the authoritative reports on the size, shape, age, and contents of the universe change too often and too radically, they have lost the cosmic imagination; they have given up judging and criticizing the march of mind, let alone taking part in it. Because matter, space, energy, causality, "laws" have acquired paradoxical or at best functional definitions, modern man has concluded that he may not *know*, and that he need not so long as he engages in operations and forecasts benefits.

All these reasons for distrusting mind combine with the absence of an accepted cosmology to rob man of his supreme pleasure and prerogative, which is to feel himself at once a moral being and a natural philosopher.* So deprived, he cannot wonder at the disease of the will and the emotions which plague him, and against which all the signs of control over nature through knowledge and machinery will not avail.

There is, to be sure, a group of scientists who give some thought to the questions of origin and form in the universe—the theorists of the "steady-state," such as Hoyle, Bondi, and McCrea. But they are handicapped by the very nature of modern science, which, no longer caring to be descriptive and particular, puts off the discussion of anything that is non-statistical. The unique defies classification, and the universe is nothing if not unique. Still, these physicists try to deal with the mystery of origins and reach conclusions that support the idea of creation.

Unfortunately, these ideas, which satisfy certain scientific thinkers, do not withstand the criticism of the philosophers. The "physics

---

* A sharp observer such as A. J. Nock expressed the common helplessness when he wrote in his memoirs: "I am far from setting myself up as a judge, . . . but when I heard of a disagreement about the shape of space, one pundit holding that space is cylindrical and another that it is globular, I went back to the eleven great theses of Pantagruel, which Rabelais says were debated 'after the manner of the Sorbonne.' . . . I turn and seek safety on the old and well-tried ground of agnosticism." *Memoirs of a Superfluous Man*, New York, 1943, 290-1.

of creation," as M. K. Munitz calls the doctrine, "not only makes no claims about the designful character of the universe; it also stops short of making any reference to the creator or the process of his making. . . . All that it would retain is the fact that matter in an elemental form is created continuously. But if the maker, the process of making, and the purpose are gone, what is there left of the concept of creation? If the sole content of the concept is . . . that matter appears or is present, then far from this being a case of creation, it is at best . . . a fact which invites scientific explanation."[14]

One could paraphrase this critique by saying that to make the process of creation intelligible, reproducible, is to declare it no creation at all; the process, or rather, the creative act, must be made imaginable and believable, which is a different thing. The same failure to grasp what is wanted explains why the mind rebels at the glibness of still other scientists who assure us that "we are just in the process of gaining a scientific picture of the total ascent of life. . . . This modern evolutionary doctrine begins with the elementary particles of the nuclear physicist and moves through the whole range of the atomic and molecular world up to the nucleic acids which, in their capacity to reproduce pattern and to pass on coded information, seem capable of forming the primitive basis for a living organism. From this point it is conceivable to move on to the gene, the chromosome, the cell, and ultimately human life. . . . This is the whole of science, engaged with a problem of majestic dimensions."[15]

If such hollow phrases delineate "ascent" and constitute "engagement," a man might be pardoned for thinking voodooism more serious and substantial than science. Fortunately, there are thoughtful scientists who refuse to deceive and be deceived by the patter of the mere advertisers of cosmology. Erwin Chargaff speaks for these skeptics when he says: "We laugh at La Mettrie's 'L'homme machine,' but 'La cellule machine' could be a contemporary title. The old anthropomorphism of the sciences has been replaced by a streamlined mechano-morphism, with some sort of templates made of stainless protoplasm. But we may ask: 'Was this change really beneficial?' "[16]

Among those who have felt mankind's need of a healing idea,

Whitehead was foremost, and he attempted to supply the want by offering a philosophy compatible with science in the great essay of forty years ago. That philosophy—metaphysics, rather—is couched in difficult language, and it moreover takes it for granted that despite the chasm between the mode of science and the experience of living, the two elements are culturally in balance. Whitehead's metaphysics therefore tries to make room for the contents of life within the categories of the space-time continuum. Thus in dealing with the person he says: "The private psychological field is merely the event considered from its own standpoint."[17] This may be a quick convenient way to define the self, but at a time when self-awareness and scientific formulation are no longer in balance at all, such statements take on the force of additional arguments for the hegemony of science. "Merely" puts the "private psychological field" in its narrow corner, and "event" gives priority to the external.

It is not my aim, even if it were within my capacity, to furnish a philosophy. My purpose remains to turn over the pieces of our scientific culture and point out their workings, which by common report are bedeviling the world. Accordingly, I reach the end of my description with the conviction I announced at the beginning, that until western man reasserts his right to be, as far as he can, a natural and moral philosopher, he will feel like an exile in his own place.

Sometimes this assertion of man's rights is itself regarded as the clue to our troubles: he worships himself instead of God, and what I suggest would seem but another step in an egocentric course. But I think I have shown how far modern man is from worshiping himself. He has given up even self-respect. If he is to climb out of his abyss, I repeat he must again philosophize. For to be a philosopher in the sense I mean is identical with being a man, and to be a man anthropos must be willing to be anthropomorphic. He can put what limitations he pleases on this indulgence, but he needs no technical authorization to feel fully himself. And though he needs none, he can find one if he insists, as he has always found what he wanted, by the exercise of mind. In this search he can be as rigorous and skeptical as any scientist, perhaps more so. For unlike the self-limited man of science, he starts with his consciousness entire, and he can say, in the words of the philosopher who distinguished between activity and process, "activity goes bail for the reality of process"; that is, we should never know there were tides if the mind were not there to ask, What is going on?

If he retreats from the indulgence of self-annihilation, man the philosopher will find ancestral voices to guide him—among them the two poets Wordsworth and Shelley, whom Whitehead named as lightbearers into the darkness that surrounds the six elements he required of a cosmology: change, value, eternal objects, endurance, organism, and interfusion.[18] But I take leave again to say that it is not six things nor sixty,* nor any combination of concepts abstracted from experience that will restore public hope: it is the passage back from *zoon* to *bios*. Man alone has a biography, and he but shares a zoology. His imagination ranges everywhere and its conflicting intuitions impel him to discover and remake the universe, never finally, never satisfactorily, but always with the exaltation of tragedy and, when no Puritanism prevents, with the gaiety of comedy. In imagination man can infer from the present universe what it was millions of years before his advent; and he can also see that it did not exist in the full sense without him; without him it is colorless, soundless, absolutely unorganized by categories of thoughts and words: as the poet said: "Earth was not Earth before her sons appeared."

It is this indispensability of man for every purpose which makes his present self-cornering in our scientific culture at once pathetic and perverse. His misplaced love of unity, which he puts in method instead of the source of the method—his mind—nullifies the advantages that brilliant method itself has gained. Instead of reserving his freedom, of being severe with techne while also loving it; of being skeptical but gay about the prospects of humankind; of being full of wonder, yet self-possessed, at the inexplicable success

---

* A French sociologist, independently of Whitehead, reminds us that man is not the abstract complex he has latterly been thought, but six kinds of selves fused into a single being: physiological, psychological, logical, metaphysical, moral, socio-political. Philippe Périer, *L'Enseignement de la Sociologie*, in *Actes du 14 Congrès International de Sociologie*, vol. III, Rome, August-September 1950, 2. In these reconstructions of man one wonders, will the glue hold?

of mathematical and scientific thought; of being aware that he—his mind, still—pre-exists all the arts and truths he plays with; finally, instead of knowing that although he is not all-in-all—no god and only intermittently godlike—he is the judge and measure of whatever haunts him; instead of these acts of intellectual courage, he prefers to think he is being driven, if not to his own destruction, then toward his ever-greater pain and confusion. He chooses to have it so; he believes like a superstitious savage in the geology of the Eocene, but he has lost the wit and serenity to see that even if he and his planet were atomized tomorrow he could and should still say today: "It will have been."

Reference Notes

Chapter 2: One Culture, Not Two
Pages 9-30

1. Max Planck, *Scientific Autobiography*, trans. F. Gaynor, New York, 1949, 33-34.
2. Arthur Balfour in *Rectorial Addresses to the University of St. Andrews, 1863-1893*, London, 1894, 319.

Chapter 3: Love and Hatred of the Machine
Pages 31-58

1. Robert Owen, *Observations on the Effect of the Manufacturing System* (1816), quoted in Max Beer, *History of British Socialism*, London, 1919-1920, I, 167.
2. Georges Friedmann, *Le Travail en Miettes*, Paris, 1956, in J. Fourastié and A. Laleuf, *Révolution à l'Ouest*, Paris, 1957, 128-129.
3. Sydney Smith, First Letter to Archbishop Singleton (1837), *Works*, London, 1859, II, 255-274.
4. Thomas Carlyle, "Signs of the Times" (1829), *Critical and Miscellaneous Essays* (Collected ed. 1869), II, 245.
5. Leigh Hunt, "What is Poetry?," *Imagination and Fancy: Selections from the English Poets* (1847), London, 1910, 60-61.
6. *John Bull*, Nov. 18, 1835, 85.
7. Sir John Herschel, "The Influence of Science on the Well-Being and Progress of Society," *Discourse on the Study of Natural Philosophy*, London, 1830, 121.
8. Irving Kristol, "The End of Ideology," *The Mid-Century*, July, 1960, 7-8.
9. Karl Otten (1920). For the parallel opinions of Carel Kapek (*R.V.R.*), George Kaiser (*Gas*), and the Expressionists generally, see Walter Sokel, *The Writer in Extremis: Expressionism in Twentieth-Century German Literature*, Stanford, 1959; and: R. Samuel and R. H. Thomas, *Expressionism in German Life, Literature, and Theatre*, Cambridge (England), 1939.
10. Henry M. Morris, "Spiritual Overtones in Engineering," *Collegiate Challenge*, Rutgers Edition, January, 1963, 16-17.
11. *New York Times*, Dec. 22, 1954. Reprinted in *Paths to the Present*, ed. Eugene Weber, New York, 1960, 394.

Chapter 4: Science the Unknown
Pages 59-83

1. *New York Times*, March 5, 1960.
2. *Science*, Feb. 22, 1963, 727-737.
3. Walter Bagehot, "Thomas Babington Macaulay," *Works*, ed. F. Morgan, Hartford, 1889, II, 58-59.
4. Jean Rostand, *Can Man Be Modified?*, London, 1959, 32-33.
5. Quoted in W. F. Monypenny and G. E. Buckle, *Life of Benjamin Disraeli*, New York, 1916, IV, 371.
6. See Victor F. Weisskopf, *Knowledge and Wonder*, New York, 1963; *Think*, October, 1960, 23-26; and *Science*, Jan. 19, 1962, 208.
7. Bentley Glass, *Science and Liberal Education*, Baton Rouge, 1959, 82.
8. Polykarp Kusch, Phi-Beta-Kappa-Sigma-Xi Lecture to the American Association for the Advancement of Science, Dec. 29, 1960.
9. Jean Rostand, *op. cit.*, 18-21; and Hermann J. Muller, "Control by Choice of Germ Cells," *Current*, January, 1962, 53.
10. Perry London in *Columbia University Forum*, Summer, 1962, 45.
11. Eli Ginzberg (ed.), *What Makes an Executive?*, New York, 1955; "What Makes a Prima Donna," *Newsweek*, February 13, 1961 (cover story).
12. *New York Times*, July 19, 1962.
13. Ernst Haeckel, *The Riddle of the Universe*, London, 1911, 380.
14. A. N. Whitehead, *Reflections*, ed. R. N. Anshen, New York, 1961, 175.
15. Erwin Chargaff, "First Steps Toward a Chemistry of Heredity," in *Essays on Nucleic Acids*, London, Amsterdam, and New York, 1963, 109.
16. Mme. Chien-Shiung Wu, on receiving the Research Corporation Award, Jan. 22, 1959.
17. Erwin Chargaff, *op. cit.*, 109.

Chapter 5: Science the Glorious Entertainment
Pages 84-118

1. Trafford, the scientist, speaking in H. G. Wells, *Marriage*, London, 1920, 502.
2. Albert Einstein, "Notes on the Origin of the General Theory of Relativity," *Essays in Science*, New York, 1934, 84.
3. A. A. Moles, *La Création scientifique*, Geneva, 1957, 106.
4. A. L. Campbell and W. Garnett, *The Life of James Clerk-Maxwell*, London, 1882, 439.
5. Albert Einstein, "On Scientific Truth," *Essays in Science*, New York, 1934/1950, 11.
6. Karl R. Popper, *The Logic of Scientific Discovery*, New York, 1961, 279.
7. J. H. Woodger, *Biological Principles*, London, 2nd impr., 1948, 217.
8. Richard Courant and Herbert E. Robbins, *What Is Mathematics?*, New York, 1941, xvii.
9. Perry Miller, "Thoreau in the Context of International Romanticism," *The New England Quarterly*, June, 1961, 158.

10. Woodger, *op. cit.*, 319.

11. C. A. Coulson, "Science and Religion," The Rede Lecture for 1954, Cambridge, 1955, 18.

12. Bertrand Russell to Sir Frederick Gowland Hopkins, quoted in James Agate, *Ego*, London, 1935, 297.

13. Woodger, *op. cit.*, 315.

14. *Ibid.*, 263. See also: Agnes Arber, *The Mind and the Eye: a Study of the Biologist's Standpoint*, Cambridge, 1954, 58.

15. Cyril Connolly, *The Unquiet Grave*, New York, 1945, 27.

16. W. H. Auden, *The New Yorker*, Nov. 17, 1962.

17. David Brewster, *Memoirs of the Life, Writings, and Discoveries of Sir Isaac Newton*, Edinburgh, 1855, II, 407.

18. Glenn Seaborg, Address to the Secondary Education Board Conference, San Francisco, April 5, 1957.

19. Walter N. Webb, *New York Times*, Aug. 8, 1962.

20. *Paradise Lost*, Bk. VIII, 11, 77-82.

21. C. H. Townes, "Some Applications of Optical and Infrared Masers," in J. R. Singer (ed.), *Advances in Quantum Electronics*, New York, 1961, 3.

22. Jean Rostand, *Bestiaire d'Amour*, New York, 1961, 125-127.

23. J. Hadamard, *The Psychology of Invention in the Mathematical Field*, Princeton, 1945, 31.

24. Bertrand Russell, "The Study of Mathematics," in *Mysticism and Logic*, London, 1949, 60.

Chapter 6: The Cult of Research and Creativity
Pages 119-142

1. David Schoenberg, quoted in "The Methods of Science," by D. Bergamini, *Think*, September, 1960, 7.

2. *New York Times*, Jan. 9, 1961; *Think*, September, 1959, 34.

3. *New York Times*, Oct. 16, 1960.

4. Horace Miner, "Researchmanship: The Feedback of Expertise," *Human Organization*, Spring, 1960, 1.

5. G. J. Renier, *History: Its Purpose and Method*, Boston, 1950, 118.

6. E. M. Forster, *Aspects of the Novel*, New York, 1927, 22.

7. See note 4.

8. Lucien R. Duchesne, *Congress Organizers' Manual*, International Congress Science Series, II, Union of International Associations, Brussels, 1961, 10.

9. J. S. Mill, in *Rectorial Addresses at the University of St. Andrews, 1863-93*, London, 1894, 27.

10. Raymond J. Seeger, "Science Education for Tomorrow," Address at Montgomery Junior College, October 20, 1960, 6 and 7.

11. Benjamin Cardozo, *Law and Literature*, New York, 1931, 7.

12. Fredson Bowers, *Textual and Literary Criticism*, Cambridge, 1959, Chap. III.

13. *New York Times*, Sept. 4, 1962; July 10, 1961; *Coronet*, July, 1959, 116; *New York Times*, June 29, 1961.

14. Herbert Read, quoted in *The Griffin*, November, 1957, 19.

15. C. E. Montague on Shakespeare, quoted in James Agate, *The Later Ego*, New York, 1951, 359.

Chapter 7: Science the Unteachable
Pages 143-163

1. G. D. Liversage, "Academic Waste," in *The Observer Weekend Review*, June 25, 1961.

2. Montgomery County [Md.] Report on Science, Washington *Post*, March 28, 1962.

3. E. L. Youmans, *The Culture Demanded by Modern Life*, Akron, 1867, 319.

4. F. A. P. Barnard, *ibid.*, 319.

5. Private communication, written for delivery at a professional symposium, Spring, 1961.

6. *Ibid.*

7. J. B. Conant, *On Understanding Science*, New Haven, 1947, 12.

8. J. S. Haldane, *Materialism*, London, 1932, 18.

9. Benjamin DeLeon, "The Scientific Approach," in *Science Learning Aids*, No. M1, Newark, 1946, n.p.

10. Editorial on Margaret Mead and Rhoda Metraux's survey of children's attitudes toward science, *Universities Quarterly*, London, November, 1957, 13.

11. Sir Harold Nicolson, "Science or the Humanities," *The Listener*, October, 1957, 697.

12. William James, *Memories and Studies*, New York, 1911, 312-313.

13. Henry Margenau, *Open Vistas*, New Haven, 1961, 201.

Chapter 8: Behavioral Science
Pages 164-190

1. Emile Durkheim, *The Rules of Sociological Method*, Glencoe, Ill., 1950, 1.

2. E. L. Thorndike, "The Nature, Purposes, and General Methods of Measurements of Educational Products," in *National Society for the Study of Education, Seventeenth Yearbook*, Bloomington, 1918, II, 16.

3. Lincoln Pettit, *How to Study and Take Exams*, New York, 1960, 19.

4. Paul Lazarsfeld, "The Obligations of the 1950 Pollster to the 1984 Historian," *Public Opinion Quarterly*, Winter, 1950-51, 636 and *passim*.

5. *Creativity and Conformity: a Problem for Organizations*, Foundation for Research on Human Behavior, Ann Arbor, 1958, *passim*.

6. *New York Times*, Oct. 3, 1962; Oct. 15, 1962; Oct. 20, 1962; Nov. 2, 1962; May 4 and May 15, 1960; Sept. 14, 1962; Dec. 1, 1957; Feb. 9, 1958; April 21, 1962; March 9 and 17, 1962; Aug. 16, 1962; July 24, 1962; Sept. 8, 1960; Dec. 10, 1960; Aug. 5, 1963.

7. Lionel Trilling, *The Liberal Imagination*, New York, 1950, 285.

8. Richard Kuhns, *Journal of Philosophy*, January, 1958, 81.

9. *Contemporary Psychology*, November, 1956, 329.

10. Karl Menninger, *A Psychiatrist's World*, New York, 1959, 468.

11. Norbert Wiener, *Cybernetics*, New York, 1948, 189-190.

12. Alburey Castell, "Education as a Goad to Philosophy," Address at the Northwest Regional Meeting of the Philosophy of Education Society, Portland [Oregon] State College, July 14-15, 1961.

13. Margaret Mary Wood, *Paths of Loneliness*, New York, 1953/1960, 183.

14. Winston Ehrmann, *Premarital Dating Behavior*, New York, 1959, 14ff.

15. Neal Gross, *et al.*, *Explorations in Role Analysis: Studies of the School Superintendency Role*, New York, 1958, 321 [and jacket description].

16. Ralf Dahrendorf, "Out of Utopia: Toward a Reorientation of Sociological Analysis," *American Journal of Sociology*, September, 1958, 123.

17. *Ibid.*

18. Hugh Sykes Davies, *The Papers of Andrew Melmoth*, New York, 1961.

19. W. Barlow, "Psychosomatic Problems in Postural Re-education," *Lancet*, 1955, 659-694; quoted in A. Friedman and H. H. Merritt (eds.), *Headache: Diagnosis and Treatment*, Philadelphia, 1959, 253.

20. C. D. Aring in Friedman and Merritt, *op. cit.*, 257-258.

21. K. Breland and M. Breland, "The Misbehavior of Organisms," *American Psychologist*, November, 1961, 681.

22. W. I. B. Beveridge, *The Art of Scientific Investigation*, London, 1955, 24, 116, and *passim*.

23. William Dement, "The Effect of Dream Deprivation," *Science*, June 10, 1960, 1705-1707.

Chapter 9: Misbehavioral Science
Pages 191-226

1. Socinus Grelot, *Sociologomachy*, Glencoe, Ill., 1961, 3.

2. Henry Margenau, *Open Vistas*, New Haven, 1961, 195.

3. *New York Times*, April 20, 1961.

4. Ben Jonson, *Timber, or Discoveries*, ed. Ralph S. Walker, Syracuse, 1953, 41.

5. Jerome Bruner, *On Knowing: Essays for the Left Hand*, Cambridge [Massachusetts], 1962, 4-5.

6. John I. Griffin, *Statistics: Methods and Applications*, New York, 1962, 49; Leonard Carmichael, *The Making of Modern Mind*, Houston, 1956, 19.

7. Lady Lovelace, quoted in John I. Griffin, *op. cit.*, 48.

8. W. Stanley Jevons, *Elementary Lessons in Logic*, London, 1870/1928, 200.

9. A. N. Whitehead, *Reflections*, ed. R. N. Anshen, New York, 1961, 176.

10. Roscoe Pound, *An Introduction to the Philosophy of Law*, New Haven, rev. ed., 1955/1959, 70.

11. *Ibid.*, 70-71.

12. Glanville Williams, *The Proof of Guilt*, London, 1958, 37.

13. See among others: F. James Davis, *et al.* (eds.), *The Sociological Study of Law*, Glencoe, Ill., 1962; and R. C. Donnelly, Joseph Goldstein, and R. D. Schwartz, *Criminal Law*, Glencoe, Ill., 1962.

14. David Abrahamsen, *The Psychology of Crime*, New York, 1960, 161.

15. Edward H. Levi, "On Legal Reasoning," in H. J. Berman, *The Nature and Functions of the Law*, Brooklyn, 1958, 328.

16. Henry Kane, *Report for a Corpse*, New York, 1948, 18.

Chapter 10: The Treason of the Artist
Pages 227-252

1. Harold Rosenberg, *The Tradition of the New*, New York, 1959.

2. J. C. Ransom, *The World's Body*, New York, 1938, 279.

3. Braque, quoted in Roman Jakobson, *Essais de Linguistique générale* Paris, n.d. [1963], 8.

4. R. Buckminster Fuller, quoted in John McHale, *R. Buckminster Fuller*, New York, 1962, 39.

5. Mark Tobey in "Threading Light," quoted in Sidney Janis, *Abstract and Surrealist Art in America*, New York, 1944, 98.

6. *New York Times*, July 13, 1963.

7. William York Tindall, *The Joyce Country*, State College, Pa., 1960.

8. Arthur Koestler, *The Age of Longing*, New York, 1951, Chap. IX.

9. Carson McCullers, *Clock Without Hands*, Boston, 1961, 90.

10. See, in addition to the daily press, *The Biology of Art* by Desmond Morris, New York, 1962.

11. Willi Appel, "Musicology," *Harvard Dictionary of Music*, Cambridge [Massachusetts], 1961, 474.

12. Leon Edel, "How to Read *The Sound and The Fury*," in *The Varieties of Literary Experience*, ed. Stanley Burnshaw, New York, 1962, 245.

13. Heinrich Zimmer, *The King and the Corpse*, ed. Joseph Campbell, New York, 1948.

14. "D. H. Lawrence and Lady Chatterley," in *Encounter*, beginning with the issue of September, 1961, 63-64.

15. George Steiner, "F. R. Leavis," *Encounter*, May, 1962, 38-39.

16. *Ibid.*, p. 37.

17. Lloyd Warner, *American Life: Dream and Reality*, rev. ed., Chicago, 1963, 54.

18. Inquiry from *Preuves* about the fate of modern abstract art, undated, but received in New York, Aug. 30, 1963.

19. Baudelaire, *L'Art Romantique*, Paris, n.d., (1927), 202.

20. Eva Lee Gallery, Great Neck, N. Y., October, 1961.

Chapter 11: The Burden of Modernity
Pages 253-281

1. Jean Nocher, *En Direct Avec Vous*, Paris, pp. 46ff. and 1960, *passim*.

2. Thomas De Quincey, *Suspiria de Profundis*, *Works*, Boston, 1873, XVI, 2.

3. Walter Pater, *Appreciations*, London, 1898, 61.

4. Leslie Fiedler, *No! In Thunder*, Boston, 1960, 7.

5. *New York Times*, May 20, 1962; C. Lévi-Strauss, *A World on the Wane*, New York, 1961; Alan Villiers, *Posted Missing*, London, 1956.

6. *New York Times*, June 11, 1963.

7. C. C. Furnas, "New Horizons in Research," Address at Brooklyn Polytechnic Institute, May 10, 1961.

8. O. P. Milne, "This England: Change in a Lifetime," *Journal of the Royal Society of Arts*, December, 1960, 11.

9. *New York Times*, Aug. 15, 1963.

10. *Selected Letters of D. H. Lawrence*, ed. Diana Trilling, New York, 1958, 215.

11. Colin Wilson, "The Writer and Publicity," *Encounter*, November, 1959, 12.

12. Autograph Letter (c. 1957-58), translated in part in *The Seventh Cooperative Catalog of the Middle Atlantic States Chapter of the Antiquarian Booksellers Association*, Item 463, 12.

13. Avner (pseud.), *Memoirs of an Assassin*, quoted in *New York Times Book Review*, Nov. 1, 1959, 35.

14. Anthony Nutting, *Lawence of Arabia*, New York, 1961, 149-150, 223.

15. Nelson Glueck in *Federal Probation*, (December, 1956), quoted in Monrad G. Paulsen and Sanford M. Kadish, *Criminal Law and Its Processes*, Boston, 1962, 161.

16. M. Esther Harding, *Psychic Energy*, New York, 1963, 139.

17. *National Observer*, July 29, 1963.

18. Catalogue of a reputable School of Nursing.

19. *New York Times*, May 18, 1963.

20. Dr. Peter J. Steincrohn, *It's Your Life—Enjoy It*, advertisement in *New York Times*, Aug. 13, 1963.

21. Robertson Davies, *A Voice From the Attic*, New York, 1960, 283.

22. *New York Times*, Aug. 6, 1962.

23. Harlow Shapley in *Charlotte* (N.C.) *Observer*, April 11, 1961, and *The View from a Distant Star*, New York, 1963, *passim*.

Chapter 12: One Mind in Many Modes
Pages 282-307

1. William James, *Psychology*, New York, 1904, I, 162.

2. Paul Vinogradoff, *Common Sense in Law*, rev. by H. G. Hanbury, 2nd ed., Oxford, 1946, 9.

3. Thomas Szasz, *New York Times Book Review* (correspondence column), June 11, 1961, 30.

4. R. W. Gerard, quoted by Michael Polanyi in "The Pioneer Beyond the Pale," by Arthur Koestler, *The Observer*, May 7, 1961.

5. D. H. Lawrence, *Psychoanalysis and the Unconscious*, New York, 1921/1960, 10.

6. Thomas Carlyle, "Signs of the Times," *Essays* (series cited), II, 251.

7. Bacon, *Novum Organum*, LXIV.

8. Wordsworth, *Lyrical Ballads*, Preface to the Second Edition.

9. Friedrich Nietzsche, *Erkenntnistheorie, Nachgelassene Werke*, Leipzig, 1904, 23.

10. A. N. Whitehead, *Science and the Modern World*, 225.

11. Ernst Cassirer, *Language and Myth*, New York, 1946, 37.

12. Bentley Glass, *Science and Liberal Education*, Baton Rouge, 1959, 115.

13. Irving L. Horowitz, *Philosophy, Science and the Sociology of Knowledge*, Springfield, Ill., 1961, 106.

14. Milton K. Munitz, *Space, Time, and Creation*, New York, 1957, 143.

15. Warren Weaver, *A Great Age for Science*, reprinted from *Goals for Americans*, Englewood Cliffs, N. J., 1960, 2.

16. Erwin Chargaff, *op. cit.*, 111.

17. A. N. Whitehead, *op. cit.*, 216.

18. *Ibid.*, 127.

Index

Of Persons and Selected Topics

Set in Linotype Caslon Old Face
Composed, printed and bound by the Haddon Craftsmen, Inc.
HARPER & ROW, PUBLISHERS, INCORPORATED